I0033250

The Art of Manufacturing

Overcome control challenges for increasing efficiency in manufacturing using real-world examples

Ninad Deshpande

Sivaram Pothukuchi

‹packt›

BIRMINGHAM—MUMBAI

The Art of Manufacturing

Copyright © 2023 Packt Publishing

All rights reserved. No part of this book may be reproduced, stored in a retrieval system, or transmitted in any form or by any means, without the prior written permission of the publisher, except in the case of brief quotations embedded in critical articles or reviews.

Every effort has been made in the preparation of this book to ensure the accuracy of the information presented. However, the information contained in this book is sold without warranty, either express or implied. Neither the author(s), nor Packt Publishing or its dealers and distributors, will be held liable for any damages caused or alleged to have been caused directly or indirectly by this book.

Packt Publishing has endeavored to provide trademark information about all of the companies and products mentioned in this book by the appropriate use of capitals. However, Packt Publishing cannot guarantee the accuracy of this information.

Group Product Manager: Rahul Nair

Publishing Product Manager: Meeta Rajani

Senior Editor: Runcil Rebello

Technical Editor: Shruthi Shetty

Copy Editor: Safis Editing

Project Coordinator: Sean Lobo

Proofreader: Safis Editing

Indexer: Sejal Dsilva

Production Designer: Joshua Misquitta

Senior Marketing Coordinator: Nimisha Dua

Marketing Coordinator: Gaurav Christian

First published: February 2023

Production reference: 1130123

Published by Packt Publishing Ltd.

Livery Place

35 Livery Street

Birmingham

B3 2PB, UK.

ISBN 978-1-80461-945-2

www.packt.com

I would like to thank Sivaram Pothukuchi for always trusting in me and giving me a plethora of opportunities throughout my professional career to learn and grow. I thank my grandfathers, Prabhakar (late) and Dattatraya Bhave (late), for being my role models, from whom I learned immensely. I also thank my parents, Suhas (late) and Tanuja, and my wife Mrunmayee, who constantly supported me in tough times and inspired me to achieve greater heights personally and professionally.

– Ninad Deshpande

Peter Gucher for giving insight into how to realize dreams. To my wife Durga for being by my side while I was dreaming and struggling to realize my dreams.

– Sivaram Pothukuchi

Foreword

Demands made on industrial automation and control are changing rapidly. Consumer demands for products are volatile and variable. Shortcomings in product quality, features, or delivery get instant publicity through social media. Online platforms increase pricing pressures. Regulatory supervision has become more stringent. All of these are demanding on manufacturers. More than anything else, there is a need for flexibility and continuous innovation. To solve these problems, manufacturers turn to industrial automation.

Industrial automation is not a single field of study. It needs a multi-disciplinary approach encompassing, of course, electronics and instrumentation. Beyond that, a good understanding of mechatronic systems is needed. Software is becoming the key factor in all kinds of automation functionalities.

Therefore the study of automation can no longer be restricted to universities. Practitioners of automation need to work on updating their knowledge continuously. Various developments are taking place - advancements are happening in the electronics of controllers and instruments; advances are taking place in communications such as 5G and also in open protocols; new methods of making goods such as additive manufacturing are coming up; the integration of artificial intelligence, personal safety, and new maintenance strategies are gaining ground. So all this has not only an implication for people who design controllers or make software programs. People who need to make specifications for automation become more and more important and face an exciting challenge.

Sustainability is an additional high responsibility for engineers. In the next 50 years, we will see dramatic changes in our living conditions. Climate change is caused by CO_2 output in heating and cooling, transport, energy production, production industries, farming, and others. You, as an engineer, can make a big difference, if you consider in all your projects to avoid these CO_2 emission factors as much as possible. Energy is a very limited good if it is not produced by burning fossil material. So energy saving in projects will be key. The shortage of other resources (aluminum, copper, lithium, etc.) will be another factor influencing production industries. As a result, each project has to be evaluated on the optimization of resources use and their recycling capabilities for the future. There will also be a shortage of clean drinking water, which is the basis of our living. So the usage of water becomes another key factor in future projects.

Wastewater handling is another responsibility of a project engineer. Take care of our rivers and lakes! Any pollution generated by a plant or factory should be cleaned up before it is released into the air that we all breathe. I hope you do not feel discouraged, but it is the responsibility of engineers to take care of all of that to save our only planet for the next generations.

In this book, Sivaram Pothukuchi and Ninad Deshpande try to present the view of automation from the application use. This, I find, is a good standpoint to take. Modern automation is largely realized by brainware. The quality of automation is determined by the quality of the engineer's thinking and software.

The essential challenges for automation have remained – to manufacture goods with higher quality, higher throughput, and fewer maintenance overheads. Since the aim of automation is to solve the main challenges of manufacturing, it makes sense to go from the problem statement to the control algorithm and jump over the details of the actual electronics.

Sivaram has been a lifelong practitioner of industrial automation and control as a developer and commissioning engineer all throughout his career, which he highlighted by a CEO position at one of the leading automation companies in India. He has a passion for imparting training and working with young people.

Ninad, too, started his career by contributing to various automation projects. Soon he was working on global standards for better interaction of different automation systems, and finally, he started his own company in the field of strategies and market intelligence.

This book will delight any curious students, practicing engineers, and laymen to get behind the scenes and understand how automation plays a role in our daily life and in the future of our planet. I can recommend this book to anyone who is involved with modern factories and machines.

Peter Gucher

Contributors

About the authors

Ninad Deshpande is a writer, storyteller, speaker, and technology evangelist with 15 years of experience in the automation industry, such as application development, marketing, and global product management.

He is an *electronics and telecommunication* engineer and holds an MBA from Symbiosis Institute of Business Management, Pune. At 30, he spearheaded the Indian marketing activities for a multinational in industrial automation. The World Marketing Congress appreciated his efforts by awarding him the *Most Influential Global Marketing Leader* award.

I thank my family for teaching me life lessons and consistently inspiring me. I admire the fighting spirit of my mother and my wife and their constant support and motivation, enabling me to achieve my dreams.

Sivaram Pothukuchi is a graduate of IIT Madras and has over 45 years of work experience in automation and control of large processes, such as power plants and power systems and also of machines in many industry verticals, such as plastics, pharma, textiles, and packaging.

He was the non-executive chairman of B&R Industrial Automation. He founded B&R India and built it to be among the top three players in the Indian market in machine and factory automation.

He believes strongly that digitalization technologies must be explored and driven into shop floor applications by young people.

I owe a debt of gratitude to my friend of many decades, Peter Gucher. Another influencer is my wife, Durga Malleswari, who has demonstrated the joy of sharing knowledge.

About the reviewer

Nilesh T. Mehta, who founded TINU TECH SOLUTIONS and has 32 years of experience in industrial automation, is actively advancing robotics and automation technology. He has worked on flatness, thickness control, and guiding systems while visiting 28 different countries worldwide.

He formerly spent 12 years in guiding, tension control, and automatic slitting systems as a director for the Maxcess International Inc. India business in the United States. In the 1990s, he developed thickness measuring with automatic thickness control (AGC), guiding systems for metal industries in India.

His professional path from a village school in Konkan, Maharashtra, to founding TINU TECH SOLUTIONS was more challenging and necessitated a significant deal of dedication to his work.

I appreciate my family and friends who have helped me along the way in my work. In today's technologically advanced world, where daily problems are evolving and new, creative paths are making life easier but more competitive. My sincere gratitude goes out to my father, who turned me into an electronics engineer and helped me get this job before he passed away in November 2022. Finally, I'm thankful to the authors for allowing me to be a part of this book on automation.

Table of Contents

3

Tension Control – Managing Material Tension   37

4

Level Control – Controlling the Level of Liquid to Avoid Drying Up or Spilling Over   53

5

Motion Control – Control, Synchronization, and Interpolation of Axes for Accuracy and Precision   67

6

Material Dispensing Control   83

Part 2: Automation and Humans   93

7

The Interplay of Humans-Machines-Automation   95

8

Automation – Dramatically Helping Avoid Human Intervention   113

9

Automation Can Build a Super-Organism with Awareness 127

10

What's Next? 139

Preface

Challenges faced in the industry are identical to those being taught in engineering textbooks. However, the way of approaching this complex challenge in the industry is different from how it is approached in textbooks. This particular gap makes fresh graduates entering the industry workplace somewhat uncertain or even apprehensive. This book aims to remove these misgivings by examining several interesting automation and control system problems in manufacturing and explains how these challenges are overcome in different real-world application contexts. The book connects these control challenges with how everyday products are made, bringing textbook solutions to life. It links textbook theory and industry practice, highlighting intricate topics related to automation and controls that need to be taught and learned.

Who this book is for

The curious layman who is intrigued by how things are made will find this book readable. This book will interest an inquisitive student of engineering (electrical, electronics, mechatronics, or Electronics and Telecommunication) who wishes to explore beyond the content of a classroom textbook. It will also serve as a teacher's handbook, helping the lecturer bring the flair of the industry to the classroom. Moreover, it will be useful for a practicing engineer, with cross-disciplinary knowledge that is needed to manufacture any real product and is not taught in engineering classes or even available in textbooks or reference books.

Here are some of the things you will learn about throughout the book:

- The role of machines, factories, and plants in manufacturing a product for daily use
- The manufacturing landscape and its continuous evolution
- Automation and its evolution – a look into the future
- Control challenges practically applied to manufacturing real-world products
- A philosophical consideration of automation
- Various applications of automation in a machine and its challenges
- Various implementation challenges
- Understanding how humans and automation work together in factories
- How the same control challenge is solved in different ways in different applications
- How terms such as Industry 3.0, 4.0, digitalization, and lean relate to each other

What this book covers

Chapter 1, Automation Is a Part of Our Daily Lives, walks the reader through elements, products, and goods that we simply take for granted, and how these goods are an outcome of automation. We will discuss how every aspect of our lives has changed as consumers have become more demanding and how this has affected the entire supply chain and the automation behind it. We will introduce the reader to a typical manufacturing setup and how machines and lines are built to enable a factory to produce the necessary products. Moreover, in this part, we will explore the control challenges associated with products across industries and try to provide insight into how automation and control help us overcome these challenges. We will start with the end product, then cover the control challenges associated, and explore the possibilities to handle these control challenges.

Chapter 2, The Art of Temperature Control, explains how managing temperature is an art in the true sense, the reason being that temperature changes are very slow and a precise way of measuring the temperature is required. Temperature is measured in manufacturing products such as bottles (plastic), plastic parts, metal, fabric, lamination (packaging), and tablet coating (in pharma). The control challenge is to maintain the temperature within the desired values (a constant temperature). These are deployed in extruder applications and bottling applications. The measurement variable, temperature, is directly measured by sensors in the heater (the heating area). In this chapter, we will read about how temperature can be controlled in the best possible way and how we can avoid overshooting, undershooting, and fluctuations. We will also explore solutions deployed in the industry such as taking measurements using a temperature sensor and PID control.

Chapter 3, Tension Control – Managing Material Tension, is all about how managing tension is essential in preparing various products such as wires and yarn. Tension ensures the diameter of yarn or wires is consistent. Similarly, there are various other industries and applications in which tension plays a crucial role. The main control challenge is to maintain constant tension or winding at a constant torque. It is also crucial for winding and unwinding materials. The measurement variable in these cases is either the direct tension of the material or, indirectly, the diameter of the rolls or the force needed for winding. In this chapter, we will look at this control challenge in detail and how products are affected if the required variables are not controlled properly. Moreover, we will look at solutions deployed in the industry to manage tension on the material, such as measuring using a load cell, making sure there is no excess tension, quickly responding if needed, or if the response time is not short enough for your requirements, then estimating the tension by measuring the diameter roll.

Chapter 4, Level Control – Controlling the Level of Liquid to Avoid Drying Up or Spilling Over, discusses the essential element in various industrial applications of controlling the fill of a tank that contains fluid. This can be witnessed in our day-to-day lives as well, with overhead water tanks that supply water at home. If the level is not maintained, then there is a risk of the tank drying up or the tank spilling. The control challenge is to manage the level of filled material (in a tank or a bottle). The measurement variable is the level, which is measured using sensors and then action is taken accordingly. In this chapter, we will look into various applications where filling containers and maintaining the levels of a contained material is crucial and how industry experts manage this control task. We will look at possible solutions, such as measuring using capacitive or ultrasonic sensors.

Chapter 5, Motion Control – Control, Synchronization, and Interpolation of Axes for Accuracy and Precision, explores how synchronization, interpolation, and motion control are vital to handling products in the industry. Machines and materials need to be precisely positioned to make a consistent and accurate product. If not configured correctly, wastage and losses occur. For making products such as bottles, wire mesh, yarn, cloth, metal parts, capsules, and syringes, motion control is deployed very frequently. The control challenge is the synchronization and interpolation of axes. The measurement variable in these applications is the movement ratio and synchronization. In this chapter, we will look at the products and control challenges in question in tandem and at the effect of not handling motion control correctly. The solution we will observe in this chapter is using drives to control motors – servos, steppers, and variable-frequency drives (VFDs).

Chapter 6, Material Dispensing Control, covers products that are filled in specific volumes, such as liquids and oils (edible oils). In these cases, filling accuracy and precision make a huge difference because if filled less, the consumer is dissatisfied, and if filled more, the company makes a loss. Thus, the control challenge is to use a fixed quantity of a filled material (in a bottle or can). In these applications, the measurement variable is the volume, making it a volumetric filling application. In this chapter, we will explore how volumetric weighing is done and how machines fill products into bottles and do so by measuring the volume. We will explore possible solutions deployed in the industry, such as measuring using a load cell for volumetric filling.

Chapter 7, The Interplay of Humans-Machines-Automation, expands on how automation was achieved in the early days by using various mechanical linkages such as wheels and pulleys or hydraulic systems for machine-based optimization. However, the actual motive power was provided by human intervention, so the machine was the means to get the movement going as per human design. Therefore, this chapter will clarify how automation provides a bridge between human design and the actual movement of a load in the industry.

Chapter 8, Automation – Dramatically Helping Avoid Human Intervention, outlines how advances in software give automation a central role. In general, an automaton is capable of performing a fixed task in a strictly limited environment. The exception is automatons that operate outside of a limited environment, and they are usually designed to reach a defined stop status and wait for human intervention. However, automation can take care of most exceptions. These are the requirements for mission-critical projects where no one failure can affect the entire system. Here, concepts such as redundant or hot-standby systems come into play.

Chapter 9, Automation Can Build a Superorganism with Awareness, teaches the reader about automation and distributed intelligence. Controller chips have been doubling in speed and halving in size every 2 years. The computational prowess of a smartphone controller is greater than the command control module from the Apollo 11 moon mission. These chips, being small, can be embedded into every device. Each of these controller boards also has the capability to communicate—as in, talk—to other controllers. This is what we mean by distributed intelligence.

Chapter 10, What's Next?, summarizes the findings of this book by explaining how humans have unique capabilities that automation will never have. It is not simply a matter of superiority. Automation has its own particular strengths, which are and will continue to be superior to the equivalent capabilities of humans. On the other hand, there are also traits that are unique to humans and set us apart. Hence, there is no place for pessimism about machines ruling the world.

To get the most out of this book

It is assumed that the reader has basic knowledge of the terms and vocabulary of electronics, electrical, and mechatronics (engineering). The reader could work in manufacturing, industrial automation, or a teaching fraternity, or be that pillar of society–the layperson.

You can try out various possible applications in the book using numerous DIY kits that are out there, but it is not really necessary. What the book teaches you is how to apply various engineering principles in real life. The best thing a student can do is join an automation company or a machine building company (an OEM) to explore these concepts further.

Download the color images

We also provide a PDF file that has color images of the screenshots and diagrams used in this book. You can download it here: `https://packt.link/2c6em`.

Conventions used

There are a number of text conventions used throughout this book.

`Code in text`: Indicates code words in text, database table names, folder names, filenames, file extensions, pathnames, dummy URLs, user input, and Twitter handles. Here is an example: "Mount the downloaded `WebStorm-10*.dmg` disk image file as another disk in your system."

Bold: Indicates a new term, an important word, or words that you see onscreen. For instance, words in menus or dialog boxes appear in **bold**. Here is an example: "Select **System info** from the **Administration** panel."

> **Tips or important notes**
> Appear like this.

Get in touch

Feedback from our readers is always welcome.

General feedback: If you have questions about any aspect of this book, email us at `customercare@packtpub.com` and mention the book title in the subject of your message.

Errata: Although we have taken every care to ensure the accuracy of our content, mistakes do happen. If you have found a mistake in this book, we would be grateful if you would report this to us. Please visit www.packtpub.com/support/errata and fill in the form.

Piracy: If you come across any illegal copies of our works in any form on the internet, we would be grateful if you would provide us with the location address or website name. Please contact us at copyright@packt.com with a link to the material.

If you are interested in becoming an author: If there is a topic that you have expertise in and you are interested in either writing or contributing to a book, please visit authors.packtpub.com

Share your thoughts

Once you've read *The Art of Manufacturing*, we'd love to hear your thoughts! Scan the QR code below to go straight to the Amazon review page for this book and share your feedback.

https://packt.link/r/1804619450

Your review is important to us and the tech community and will help us make sure we're delivering excellent quality content.

Download a free PDF copy of this book

Thanks for purchasing this book!

Do you like to read on the go but are unable to carry your print books everywhere?

Is your eBook purchase not compatible with the device of your choice?

Don't worry, now with every Packt book you get a DRM-free PDF version of that book at no cost.

Read anywhere, any place, on any device. Search, copy, and paste code from your favorite technical books directly into your application.

The perks don't stop there, you can get exclusive access to discounts, newsletters, and great free content in your inbox daily!

Follow these simple steps to get the benefits:

1.   Scan the QR code or visit the link below:

https://packt.link/free-ebook/9781804619452

2.   Submit your proof of purchase

That's it! We'll send your free PDF and other benefits to your email directly.

Part 1:
Introduction to the Manufacturing Landscape and Innovative Automation in Everyday Life

By the end of this section, you will understand the manufacturing landscape, supply chains, and basic elements of a factory and a machine, moving from larger parts such as a factory and machine to the smallest element such as a sensor or an actuator. We will also explore the topics of control challenges across a selection of industries and try to provide insights into how automation and control helps us overcome these challenges.

This part has the following chapters:

1
Automation Is a Part of Our Daily Lives

An attempt to describe automation can be like the five blind men in the proverbial story encountering an elephant for the first time – each explores one aspect of the animal and narrates his experience accurately, yet all the narratives together do not yield an overall picture. We can look at automation in the following ways:

- Automation is a subject that can be examined in terms of components

- Automation can be examined in terms of the system software, application software, and libraries

- Significantly, automation can be understood by understanding the algorithms and the products

When we stack up these views, we are able to build a realistic 3D image of automation.

Control challenges in the production of objects of everyday use are many. A close look at the kind of challenges that are encountered opens an interesting puzzle. Problem categories are few because the object parameters that need to be controlled are few. These could be, for example, the physical dimensions of the object, weight, thickness, and surface finish. Corresponding control variables are nearly always the same—speed, rate, tension, position, thickness, pressure, level, temperature, and so on. Control algorithms themselves are notably limited, for example, proportional control, integral control, and **proportional, integral, derivative** (**PID**) control. What is the puzzle that we mentioned at the beginning of the paragraph? The puzzle involves so many of the challenges that arise during implementation in an industry situation. These challenges arise from less-than-ideal measurements, control responses, imperfect earthing, and other problems. In spite of all these shortcomings, the machine must function properly and deliver good products. This is the magic achieved by clever control algorithms. Control schemes and algorithms are the solutions to the puzzle.

This book is meant for students but even more for faculty. There is always a gap between learning theory from textbooks or classrooms and learning from a hands-on lab experiment. A lab experiment is oriented to deepening your understanding of the theory. Too frequently, we encounter the opinion that academic design is not useful in industry practice, and that there is a gap between the lab and the shop floor. This book express the views of the industry and hopes to bridge this gap. Each chapter can be a starting point for creating a lab setup for investigating actual issues in industrialized automated machines. We do not delve into theoretical aspects of control loops and control design. The storyline is from an industrial implementation point of view.

Moving on to the focus of this chapter, would you be surprised if I told you that we are surrounded by elements, products, or goods that are an outcome of automation? Manufacturing is the backbone for producing these everyday products that we so often take for granted. As an engineer or a technical person, you learn so many aspects of engineering but can sometimes fail to understand how these translate or are applied to manufacturing products. We will introduce you to a typical manufacturing setup and demonstrate how machines and lines are built to enable a factory to produce the necessary products. We will also introduce the basic elements of a factory and manufacturing line. In addition, we will also introduce the aspects of how automation, control systems, robotics, and mechanics are instrumental in building up a machine. This will help you understand the finer elements of an entire manufacturing setup. You will be able to envision this setup, which forms the basis of the book, and also start relating various daily used products with manufacturing and automation setups. We will take you through the automation components and structures, helping you understand these elements in detail that will also be carried forward throughout the book.

This chapter will cover the following main topics:

- Understanding our daily rituals and the products we use
- Probing how consumer demands and needs are drivers
- Analyzing the elements of a factory – production lines and machines
- Analyzing the elements of automation – components and structures

Understanding our daily rituals and the products we use

"What is going on?" wonders Jacob on a rainy afternoon while returning from his final day at his internship. Jacob is a brilliant third-year electronics and telecommunication engineering student who got an early opportunity to work as an intern during his semester break. He has been getting excellent grades during his previous semesters, but his internship has been an eye-opener. He has realized that even after being fantastic in his studies, his time in the industry has been filled with hurdles. Being a sincere student, he feels dejected due to the vast gap that exists between his curriculum and the expectations in the industry. That evening over dinner, Jacob is quiet, and his mother realizes something is wrong and decides to talk with him. At first, Jacob doesn't feel comfortable enough to open up to her, but with her persistence, he explains what is on his mind. Being a banking professional, she does not understand what he is trying to express, and she tries her best to motivate him. However, Jacob is still unconvinced and uncertain about many topics. His father, a senior and respected professional in automation, overhears this conversation between his wife and his son. He is instantly taken back to the days of his first job, where after graduating in electronics engineering, he was exposed to the world of automation. He instantly recollects many facets of what his son is going through and realizes

that he needs to do something so that his son starts believing in engineering studies and how the industry operates. He faced a torrid time during his initial work years as he was unable to understand that even after being a high achiever in college, the activities in the industry were way too different, and the industry had not only varied but also very high expectations.

He decides that he will pass on all his knowledge to his son so that he is able to overcome these challenges without too many difficulties. He decides to have a detailed discussion with Jacob.

The following day after breakfast, Jacob realizes that his father is trying to get into a conversation. He understands that it must be the continuation of his conversation with his mother, and he is not wrong.

Both of them enjoy a long conversation about how automation is central to what he is studying and the career he will choose.

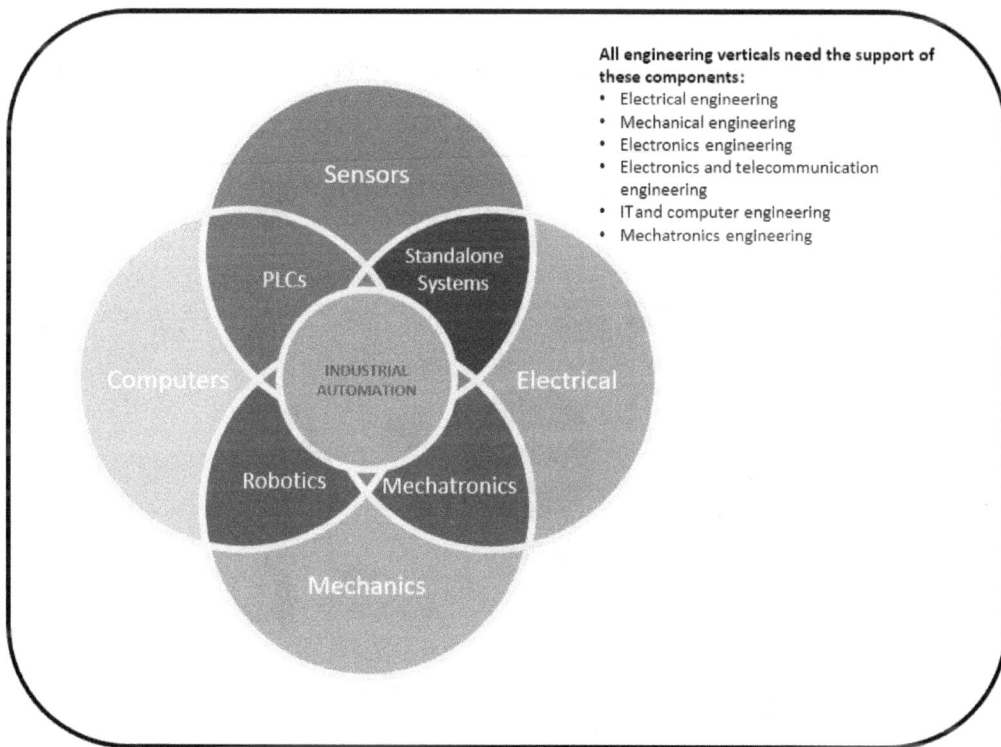

All engineering verticals need the support of these components:
- Electrical engineering
- Mechanical engineering
- Electronics engineering
- Electronics and telecommunication engineering
- IT and computer engineering
- Mechatronics engineering

Figure 1.1 – Engineering disciplines have these fundamental elements

We will take you through their conversation in the following chapters.

How do you start your day? You wake up and most probably head to the bathroom to brush your teeth. You pick up the toothbrush and apply the toothpaste to it. You might also own an electronic toothbrush as well as a plastic cover to protect or keep your toothbrush bristles safe and clean.

Heading to the breakfast table, you serve yourself delicious food. In addition, you use cutlery and plates to eat your food with. Having a bath and getting ready, you need a block of soap, body wash, shampoo, conditioner, and a towel. You then open your closet and take out clothes. You like to utilize the refrigerator to make your packaged milk last longer. You also keep various food and beverages, such as butter, cottage cheese, meat, cheese, ketchup, ice creams, bread, and so many other things stocked up in your refrigerator. Heading out of the house for work or college, you either drive a car, motorbike, or bicycle, or take the subway, train, or bus. At work or college, you open a book and write in it with a pen, and you also use various stationery at your workplace. If you feel thirsty or hungry, you visit the store and pick up a bottle of water, a packet of chips, or packaged food. You might even wish to buy a soft drink. At lunch, you open your lunch box and enjoy your home-cooked meal. In the evening, you go out for a cup of tea or coffee. On coming home, you either open your laptop or turn on your smart TV to watch some web series or sports. You also load your washing machine and start the cleaning process. You might have a robotic cleaner to help with the sweeping and mopping. You then make your bed by removing the bed cover, putting on bed sheets and pillow covers, and setting your comforter. You might even put on your air conditioner before getting a good night's rest.

Figure 1.2 shows many of the common devices and foods that you can use on waking up at home:

Figure 1.2 – Some of the devices and foods that we handle on waking up at home

All these products mentioned here are just the tip of the iceberg of the enormous list of products that are the outcome of automation. Surprised? We are not joking; all these products are made in a factory following a process on different machines, maybe even different locations. In addition, the supply chain, logistics, and distribution play a phenomenal role in not only manufacturing the product but also enabling the organization to make it reach its desired destination and, eventually, its consumer.

Let us take the example of a supermarket and identify various products that we see on the shelves. In the vegetable corner, you find some fruit or vegetables that are wrapped in a thin film and sealed. Film wrapping is very common in the packaging industry to enable foods to be contamination-free and provide safe transit. These concepts are also applied to bottles, soda, soft drinks (bottles and cans), and oil cans, enabling easier handling, transport, and stocking. You have likely observed such packaging in many stores and supermarkets. *Figure 1.3* shows how cans or bottles are shrink-wrapped in factories:

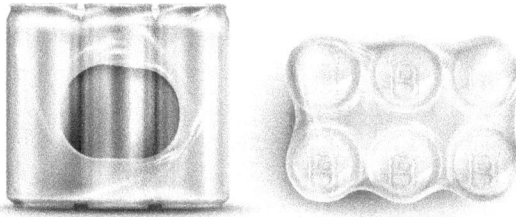

Figure 1.3 – Shrink-wrapped cans in factories

In addition, filling liquid in these bottles is also a very different kind of automation that requires accuracy and precision. We will walk you through these processes in further chapters. Ideally, the bottling line will incorporate sections such as blowing the bottle, filling the bottle with the desired fluid, labeling the bottle, and then capping it. When we reach the aisle with all the packaged foods as well as consumer and personal care products, such as lentils, flour, pulses, soup, ready-to-eat items, shampoo sachets, biscuits, cookies, chips and chocolates, we see perfectly packaged and weighed packets. These packets are also an outcome of complex machines and automation. **Vertical form fill seal (VFFS)** and **horizontal form fill seal (HFFS)** are the machines that are deployed in the factory for producing such products. These packaged products are shown in *Figure 1.4*:

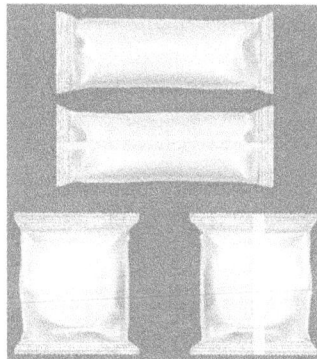

Figure 1.4 – Packaged products that are manufactured on VFFS and HFFS machines in the factory

I would go one step further and say that all the products in the supermarket are made using some form of automation. Until now, we have just focused on the food and beverage and packaging industries. However, there are so many industries, such as automotive, textile, plastics, printing, and pharmaceuticals, that also have immense automation. We will touch upon the manufacturing landscape and automation capabilities in all these industries in the coming chapters.

By this time, you will have realized that everything that surrounds you is, in one way or another, related to automation and manufacturing.

Probing how consumer demands and needs are drivers

The market has moved from a supplier market to a consumer market. All organizations have realized that customer-centric behavior enables higher growth and improves customer loyalty. Even for organizations working in the **Business-to-Business** (**B2B**) domain, it is not that different. An organization in the B2B domain is one where the organization directly sells goods and products to another organization and not to the consumer. Let us take a simple example of buying a laptop. An individual can buy a laptop online or from the store irrespective of the brand. Thus, companies such as Dell, HP, etc., would be an organization in the **Business-to-Consumer** (**B2C**) segment. However, electronic chips are needed to manufacture laptops, which might be delivered by a company like Intel. Therefore, Intel will sell its chips to HP, Dell, etc. Thus, Intel would work in the B2B domain. In most organizations, you will encounter the statement *the customer is king*. This is largely evident in the way products are made and offered to us today versus a few decades back. You must have seen that over the years, products have undergone a huge change. In the 1990s, food items were sold loose; however, due to hygiene requirements, ease of transportation, and storage, you mostly find products that are well packaged or you prefer buying packaged products. What has led to this change? As you might rightly point out, consumer and market demands are acting as drivers. There is a similar situation in the automotive industry. In the 1980s and 1990s, for instance, there were hardly any vehicles and variants in the Indian market, which was mostly dominated by a few manufacturers. Today, you have local and multinational brands offering you multiple variants. In simple terms, you are spoilt for choice.

All these changes are solely due to consumers demanding the supplier provide these possibilities. Today, it is just not enough for you to own a car or a motorbike; equally important are the features being offered and the cost-to-feature ratio. It is also possible that you might even pay a little more for that additional comfort.

Consumer awareness has increased. As a result, consumer demands on quantity, quality, safety, and price have all gone up. It can be very difficult to satisfy this need without automation.

Analyzing the elements of a factory – production lines and machines

By now, you might have a fair idea about how everyday products are made and eventually how to associate this with the manufacturing landscape and automation. Let us dive deeper into the concepts of the factory, a line, and a machine. I am sure these terminologies must be totally new for many of our readers. A manufacturing setup is primarily split into two main areas, the **information technology**

(**IT**) infrastructure and the **operational technology** (**OT**) infrastructure. As the name suggests, the IT infrastructure in the factory is responsible for handling all the IT activities and managing all data. The production and related automation take place in the OT area. The factory floor is the place where all the machines are lined up in a particular sequence to manufacture a particular product.

Continuous process versus discrete process

There are two types of processes in manufacturing – process automation and discrete automation. Process automation can also be termed as continuous process automation; as indicated by the name, it is a continuously operated process. These specific automation possibilities can be together or in separate areas. A **continuous process** is a process where the end product or finished product is manufactured continuously without a break, and the manufacturing demands this continuous activity. Wherever there are process-based operations, any untimely stoppages usually lead to wastage and losses. Moreover, restarting the process takes a lot of time. It is important to understand that the continuous process is well connected, and an error in any area disrupts the entire process. Thus, these are critical operations and need critical infrastructure for handling such processes.

On the other hand, **discrete processes** are independent operations where there are links but then any untimely stoppages in one area do not necessarily directly affect the other areas. However, stoppages in discrete processes usually lead to bottlenecks in the system. With discrete processes come the concept of a line and machines. Discrete processes are built up of several machines that, in turn, are connected with one another to form a line. Each machine is responsible for a particular activity or process. A product must move through these machines and undergo processing at each station. Thus, raw material is fed into one end of the line, and a finished product is obtained at the other end. Let us take the example of a soap-making plant.

Soap making combines both areas of continuous process and discrete process. Preparing the soap mixture is a continuous manufacturing process. This forms the raw material for manufacturing the actual soap block that you can find in supermarkets. A brick of soap is first split into smaller soap blocks at the first station. Then these blocks get punched with the brand or the name of the soap at the following station. After this, the soaps are wrapped in branded glossy or carton packaging. There could be multiple stations for wrapping the soaps. A conveyor takes the soap block through each station. At the end of the wrapping line, the soaps are put together in a huge carton and then shipped to the warehouse for further distribution.

Mass/batch production versus customized production

The process described earlier is a form of **mass production** or **batch production**. Thus, once the batch is started, the batch will only finish after the raw material is used up. Let us again head to the kitchen to check an analogy. If you have one burner and are already preparing tea, if someone asks you to prepare coffee, you will need to first finish preparing tea. Once you are done preparing tea, only then will you be able to prepare coffee. This is an example of a batch process, where you need to finish one batch before you start a new or a different one. Thus, on a single machine or a line, there could be multiple soaps that might be manufactured. However, before you take up manufacturing a soap of a different scent, you need to finish the batch at hand. You might also need to clean the existing equipment to avoid contamination.

This is a concept that was introduced in automotive manufacturing. On the other hand, **individualized production** is a relatively new concept where each product can be customized. Today, you can gift chocolates with personalized messages and individual names. You need to pay a premium to purchase such products, but there is immense pleasure in having or gifting a bar of personalized chocolate. With adaptions made to existing production lines, organizations enable such unique possibilities. The major focus for organizations is to have a perfect balance between the cost of producing individualized products and not overburdening the consumer. Simply put, the manufacturing cost cannot be too high and the unique product cannot cost too much. Let us follow an example to better understand this: if a bar of chocolate costs $10 with a manufacturing cost of around $2, then the cost of personalized chocolates cannot be $20 with a manufacturing cost of $12. There will always be a limit on how much premium a consumer is willing to pay for a special product. Thus, the focus for organizations is to keep the costs under control while also offering such unique products that bring something new to the market.

Batch production or mass production provides an organization with cost-effective manufacturing. Moreover, this kind of production methodology is developed to cater to rising demands as well as keep costs in check. Once again, the focus of these innovations and changes is the consumer. As the consumer demands such possibilities, organizations are working toward building such products. In particular, to satisfy the pressure on price, goods need to be manufactured in large quantities at high speed. This is a case for automation.

Inside a factory

Every factory in every industry is a bit different; however, the basics always remain the same. In a factory, there is always an area for the corporate teams that might include research and development, human resources, finance, marketing, management, supply chain, and sourcing, among many others. This forms the office space of a factory. Then there is an area where the goods are manufactured, such as the shop floor. This area forms the OT space inside a factory. There are other areas, such as stocking incoming raw materials, stocking finished goods, the warehouse, and the distribution of products. In larger factories, there could be a possibility that these elements are not in a single area but might be spread over acres of land or even across one city or multiple cities.

There are even factories that are set up in different parts of the country or even the world, while still being connected to each other. There are examples of factories in the US, Europe, China, and India that are connected via the internet for various operations. These are the cases for huge organizations that are spread globally.

Let us now focus on one part of the factory that is of utmost importance for this book—the shop floor.

The **shop floor** or the **factory floor** is the place where the products are manufactured. The question is how? The shop floor is made up of different machines. Each machine is responsible for performing a certain action. Let us take the example of preparing a burger in any big burger shop. The bun needs to be toasted, which happens on, say, counter no. 1. A person manning this counter will toast the buns and, once done, will pass on these buns to the next counter. The buns then need to have some sauces applied to them, which takes place on counter no. 2. The person manning the counter then applies sauces on the buns based on the order. The buns also might need the addition of lettuce, which can take place either on a subsequent counter or on the same counter. After this process, this product is moved to the subsequent counter, say counter no. 3, where a deep-fried patty is inserted between the buns. Finally, the burger is ready and then this product is moved to another counter, say counter no. 4, where the burger is packed in a paper carton and sent to the delivery counter. Thus, the counters for assembling the burger are stationary and the product, that is, the burger, makes its way from the first counter to the final delivery to the consumer. There is also a possibility that along with the product, the person manning the stations moves with it or there are different people manning different counters. On the shop floor, these individual counters are called stations and are made up of individual machines. Each machine has a unique role to play. These stations together are termed as a line in the factory. This forms the basis of a machine and a line. A typical line in the packaging industry is shown in *Figure 1.5*:

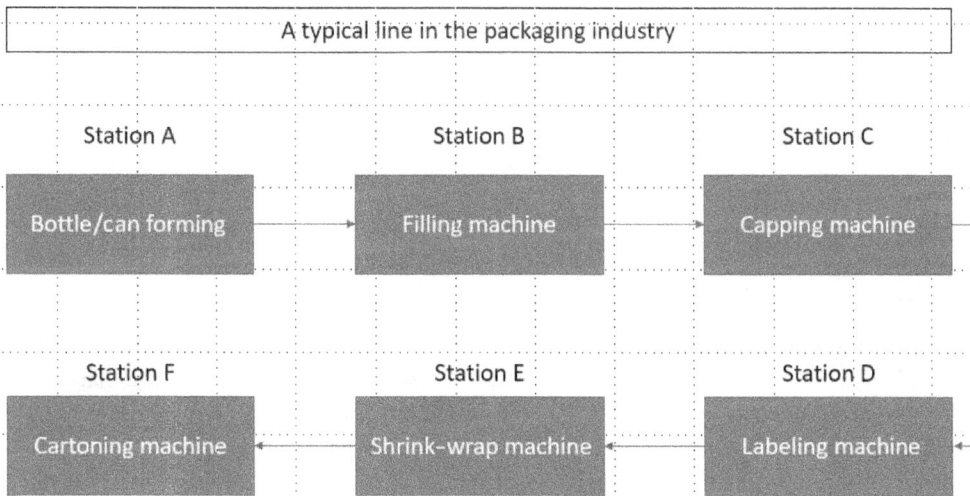

Figure 1.5 – A typical production line in the packaging industry

Figure 1.6 shows a typical production line with various stations, robots, operators, conveyors, and products:

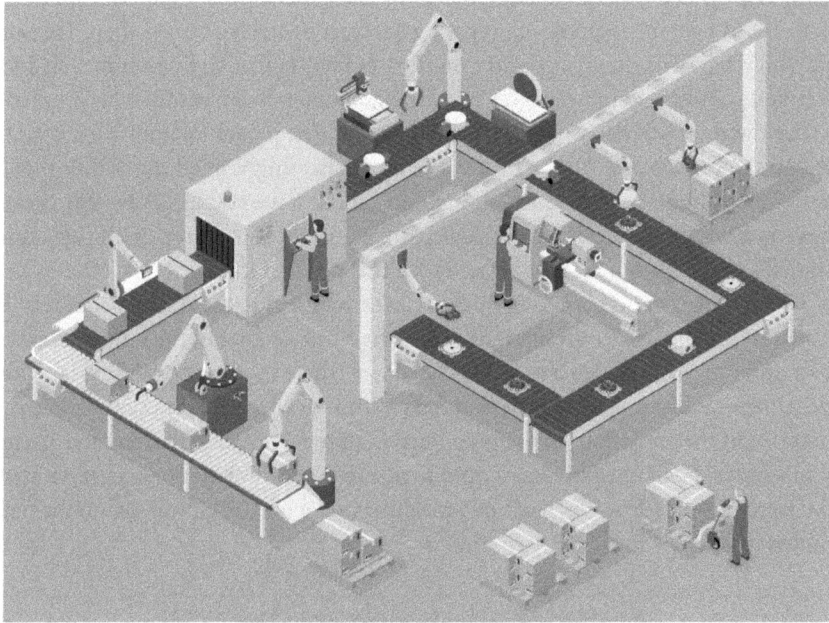

Figure 1.6 – A view of a production line with various stations, operators, and conveyors

An established method, inherited from the second Industrial Revolution, is to divide work on a product. This work or process is repeated on each product at one place called a workstation.

Let us take the example of a factory in the packaging industry to clarify how these concepts add up. In the packaging industry, let us focus on a bottling line where the factory manufactures and fills water bottles.

Water is one of the raw materials, and the factory needs to have a huge storage of water for filling the bottles. The PET bottles are made from a miniature plastic preform that is no more than 10 cm in length. This is another raw material for the bottling factory. This preform is molded into a PET bottle with a capacity of 1 liter using a process called PET blow/blow molding. This is the first station in the factory, where the preforms are blown into the desired shape of a bottle. The bottles are then transferred either manually or on a conveyor and taken to the subsequent station. At this station, the bottles are cleaned and then filled with water. These filled bottles are then moved on to the next station on the same conveyor for capping. After applying caps, the bottles then have labels applied for various purposes, such as branding, compliance assurance, and other information, and then are checked for quality. If the bottles are unequally filled or there are quality issues with the bottles, capping, or labeling, then this station rejects the faulty bottles and only accepts good bottles. After this, the conveyor takes the bottles to either a carton or shrink-wrap station, where the bottles are packaged for transport. A robot then creates a stack of these packaged cartons or shrink-wrapped bottles and they are made ready for storage or dispatch. The conveyor is responsible for transferring the products from one station to another. The stations are individual machines and are controlled by individual controllers, which we will see in the next section.

The line on the factory shop floor is a network of many machines, one after the other, performing different actions in order to manufacture a complete product. A machine is an independent entity that is responsible for performing only one or a set of tasks. Thus, any faults in one machine do not affect the operation of one on another station. However, as discussed earlier, if one station is down, then it leads to a bottleneck, as the previous stations in the line are healthy and producing at full speed, and the stations after the faulty station are healthy but are starved of products, as the station feeding them with products is broken down. Thus, as soon as the machine with a fault is back up and running, the line immediately starts operating at full speed. Systems resuming with a healthy status and returning the entire line in a factory to full speed operation is a scenario that is usually possible in discrete manufacturing with machines and lines; however, this is unlikely in a continuous process.

It is possible that inside a factory there could be multiple lines and tens to hundreds of machines on each line. There is the possibility that an entire building is dedicated to one line and there might be multiple buildings/shopfloors that constitute a factory. It is also possible that in one building there could be multiple lines for different products. In a printing press, there could be four lines printing different newspapers of the publication in different languages. In the case of a packaging line, there could be three parallel lines that are manufacturing, packaging, and cartoning different types of cookies/biscuits of the same brand. In the case of an automotive factory, there might be different buildings for the body shop, the paint shop, the engine line, the chassis line, and the final assembly, with each line having different machines and robots.

As we are now familiar with the elements of the factory, lines, and machines, let us go a step further and understand the finer elements of a machine.

Analyzing the elements of automation – components and structures

Viewing automation in terms of its components can be linked to the study of the morphology of living things. These components, at a broad level, are the controller, sensors, and actuators.

A machine is a mix of mechanics, mechatronics, electricals, and electronic components. The mechanical components are the gears, conveyors, structure, safety guards, base or foundation, hydraulics, or pneumatic connection. The electrical installations usually cover the electrical panels where all the electricals and electronics are mounted, such as the cabling and power connection. The electronics are components such as the **programmable logic controller** (**PLC**), **human-machine interface** (**HMI**), drives, motors, **input/output** (**I/O**) modules, sensors, actuators, and programming software. All these together form the machine. Without even a single element, the machine will fail to operate. Each element needs detailed attention, and this is the way any machine is built, developed, and then commissioned on any factory floor. The traditional approach to machine building involves building mechanical components as it is the most time-consuming and resource-intensive activity. Once the mechanical element is nearing completion, the electrical cabinet manufacturing and cabling can be started as it is the next most time- and resource-intensive work. After this, the software can be developed, installed, and tested to check the entire machine. However, with various advancements such as digital twins, the simulation of the electrical and software development can start in parallel, helping machine builders achieve a reduced time to market.

Control, automation, and data technologies

Automation forms a part of Industry 3.0, and can also be called **digitization**. Data technologies, which are the backbone of Industry 4.0, provide the means to monitor and assist managers to make competent decisions, based on events and measurements from the shop floor.

Control is the design effort to keep a chosen parameter within defined limits, even as the process environment changes. Control predates digitization and even electronic controls. Automation is a means of defining in great detail a control process and a means to perform the control task tirelessly and repeatedly according to the programming.

Automation is not technically a necessary element for data technologies. Yet, if the elementary parameters of throughput and quality are not achieved using good control automation, the benefits of data techniques cannot be obtained.

Now, let us understand the control loop. There are three standard parameters in a control loop. There is, first and foremost, the parameter to be controlled; it is called the **control variable**. It could be the level of liquid in a tank, the temperature of a substance, or something along similar lines. Next, there is a **set point**. This is the desired value to be attained or maintained by the control variable. There will be a measurement of the actual value of the control variable—what its value is now. The difference between the set point and control value is the deviation, usually called the **delta**, which is the third parameter. This deviation will act on a control element—a valve or a heater or something—to reduce this deviation. We will explore many control schemes. In every chapter, we will take objects that you encounter in everyday life. We will touch briefly on some of the processes involved in the manufacturing of these items. We will sketch out the control schemes employed to improve productivity in this manufacture. We will give some hints as a starting point for faculty and students to devise experiments in controls using automation:

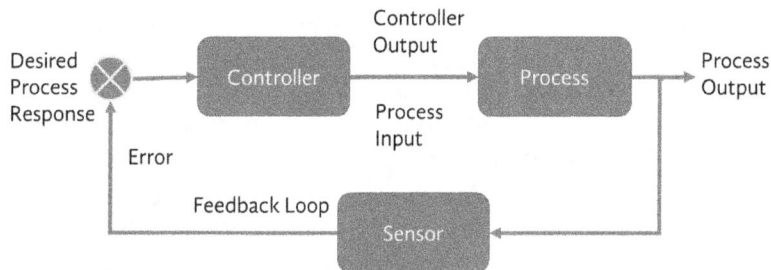

Figure 1.7 – A typical closed-loop system in industrial automation

An efficient control algorithm is one that keeps the **process variable** (**PV**) within permitted limits around the set point. The greaterore the number of excursions, the higher the amplitude of the excursions, and the worse the algorithm is. Some of the factors that influence the behavior of a control algorithm are the inertia of the load system, noise in measurement, the frequency of sampling of the control variable, the rate of change of the process variable (for example, jerk change), and the rate of execution of the control program itself.

Control algorithms and PLC programs

A PLC program is a representation of the control algorithm in software. A typical PLC program cycle works like this: initially, all inputs are scanned and recorded in the local memory. This constitutes the input image and is assumed consistent; that is, all values have the same timestamp. Then the logic is executed, and outputs are calculated and written to the output image. Finally, the output image is transferred to the actuators. Hence you can see that there is, in the worst case, a latency of three cycle times. The discussion becomes more complex if you consider remote I/O systems, multiple controllers communicating on a bus, and what is very common—control loops inside control loops. The inertia of the load is not always mechanical inertia in moving parts. It is sometimes, for example, thermal inertia, such as when you try to heat an object, depending on its heat capacity, the response (increase in temperature) can be slow. The control algorithm needs to have this inertia and the latencies as factors. Control algorithms need tuning for a given installation. This is mostly a manual input and is provided by experienced operators. Present-day practice is moving toward auto-tuning to increase operator comfort. A practical issue is that during tuning, the material is wasted. Hence, the need arises to shorten the time needed for tuning and the trial-and-error involved so that wastage is minimized.

We are primarily going to examine automation in the field of manufacturing; therefore, we always talk of industrial automation. Thereby, we distinguish it from office automation—by which we refer to the hardware and software used for the automation of office functions, such as accounts, HR, sales, and inventory. In this book, we will walk our way through using control algorithms as our street guide and understand them by exploring how they play a role in making some of the very familiar products we use daily. This structure allows us to have a good idea of the progress achieved as we go along. The idea is that some control problems are common to the manufacturing of many products, and at the same time, any product will use many control elements. It is also very interesting that the same control problem is solved using different control elements, different control algorithms, and, particularly, different control strategies. Industrial automation is a complex organism. It has a structure and a function. There is a logic that drives the function and the structure. The structure evolves, and the development is guided by the intended or foreseen functions. Sometimes it works the other way as well; because of developments happening in components, some devices get developed. Then, new, innovative applications are created and put out on the market.

Tools and machines

A tool is a device that enhances the strength of a human. It mostly needs a human to wield it. There are also machine tools, where the machine uses the tool to achieve desired modifications in the workpiece. Examples of simple tools are the hammer, saw, and screwdriver.

A machine in our context is an apparatus (a mechanism) that converts a workpiece from one shape to another, which is either ready for dispatch or ready to be further processed by the next machine.

The drive for more industrial automation

Modern industry needs more flexibility. Production volumes need to be scaled up or down. The product mix needs tuning regularly. Rules and regulations change. All this also calls for automation.

Automation components

We can list the components moving from the smallest element to the largest—sensors/actuators, I/Os, drives and motors, the controller (PLC), and the software to program the PLC. All these elements have to be compatible and should be able to communicate with each other so that they work as desired. But you can see that each element has a different function and design. So, it is important to note that the sum of these components is much more capable than any one individual part.

Programmable logic controller (PLC)

The PLC is the brain of the automation system. As the name suggests, the PLC is an electronic component that can be programmed as per the needs of the user. The program can be written, deleted, or re-written in the PLC. The data from the machine operations is also written and stored in the PLC. The PLC is identical to your computer or laptop. It has a processing chip, a motherboard, **random access memory (RAM)**, **read-only memory (ROM)**, a clock, storage space, Ethernet ports, and USB ports. In the industry, it can either be a microprocessor-based, **field-programmable gate array (FPGA)** or an industrial PC:

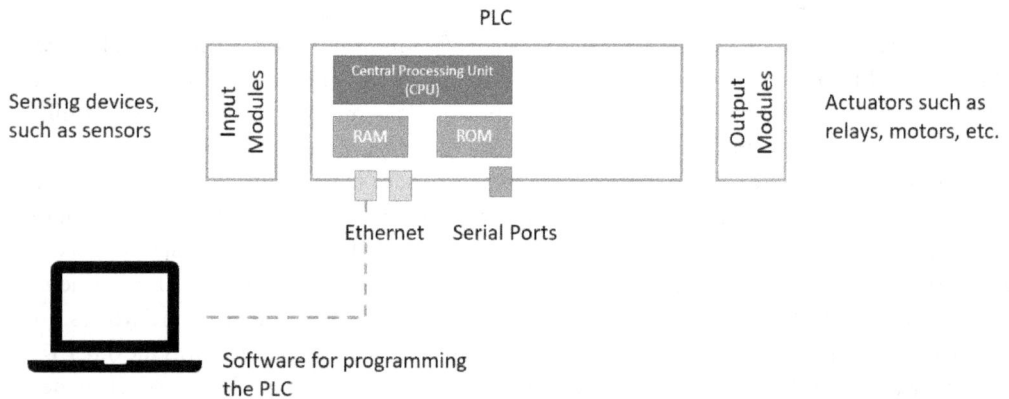

Figure 1.8 – An architecture of the PLC with hardware and software functions

Software is needed to program these PLCs, which we will see in the upcoming subsection, *Programming software and tools*. This software is installed on a laptop or a desktop, and the developer then connects their laptop with the PLC for programming it.

Inputs and outputs (I/Os)

Just like our hands, ears, nose, tongue, and eyes are the sensory organs and provide the necessary information to our brain, some inputs are required for the PLC to function efficiently and take necessary actions. When we pick up something that is very hot, our fingers realize it and send signals to our brain, to which our brain responds by most probably dropping the object we picked up and holding our hand in cold water. Thus, the input from our finger is translated into action.

Similarly, the PLC needs inputs from various devices, such as sensors, to gather information about what is happening in the machine to take necessary action. I/Os in automation are extremely essential elements. As we saw earlier, inputs are needed to gather data from the field from sensing devices, and outputs are needed to take action based on the inputs received from the field. The software inside the PLC converts the raw data from the inputs into actionable information. Let us take an example of a tank being filled with water having two sensors for sensing the levels—one at the bottom and another at the top. When the sensor at the bottom is not triggered (sensed), then normally, the tank is empty, and water should be filled into the tank. When the sensor on the top is triggered, that means the tank is full, and the water filling should stop and remain stopped until the sensor at the bottom is once again not triggered. However, as a programmer, you need to take care of faulty sensors. What would happen if the sensor at the bottom is faulty and shows an indication that it has not been triggered , but the tank is full? The programmer needs to understand these possibilities and program accordingly. In addition, apart from these sensors, the market also provides analog ultrasonic sensors for measuring levels. We will study these in detail in *Chapter 4, Level Control: Controlling the Level of Liquid to Avoid Drying Up or Spilling Over*.

Typically, signals are ideally either digital or analog. A state of 0 or 1 represents *off* and *on*, respectively. A digital input, as well as digital output, has the same binary representation of 0 and 1. When you need to start a machine and you press a button, the input 'on' (binary 1) is sent to the PLC. This is a momentary push of a button. The moment the button is released, the signal changes to 'off' (binary 0). Similarly, to start an operation, a PLC needs to send an 'on' (binary 1) signal to the device and if you need to stop the device, then an 'off' (binary 0) needs to be sent to the device.

On the other hand, analog signals have a range that might be 0 to 65,535 or –32,768 to 32,767 and can be scaled to identify the actual value. Typically, analog signals are used for frequency, pressure, temperature, and flow.

Sensors and actuators

As described in the *Inputs and outputs (I/Os)* subsection, the sensors are connected to the inputs and the actuators are connected to the outputs of the PLC. Sensors are, let us say, the eyes and ears of the controller, whereas actuators are the arms and legs. Similar to our sensory organs, the sensors sense changes in the machine and provide them as inputs to the PLC. The PLC, in turn, runs the program and decides what the predefined actions are that are needed after particular inputs are activated. Let us again take the example of the water tank. What is the meaning of filling water? In most common situations, water would be filled into the tank by either turning on a valve or switching on a motor. These are outcomes of switching on outputs from the PLC and physically connecting them to the motor or the value activating them. When the motor or valve is switched on, there could be feedback to check whether they are actually turned on. These are all various possibilities, and the programmer needs to take care of these aspects.

There are many types of sensors, such as capacitive and inductive, that provide digital signals to the PLC and need digital inputs for detecting these inputs. There are also sensors that are analog in nature; that is, they provide inputs in the value range of 0 to 65,535 and need an analog input to sense these signals. There are also sensors that provide pulse-width-modulated signals, and some are high-speed inputs that need special input modules. There is a large set of potential inputs, and we have only covered some basics in this section.

Similarly, there are different types of outputs for various actuators, such as a digital output providing signals in 0 or 1 for actuating the output. There are relay-type outputs too. Outputs also come in the form of analog outputs where the activation of the outputs varies from 0 to 65,535. There are pulse-width-modulated outputs as well as high-speed outputs.

The following is a list of some of the sensors used in automation:

- **Inertia measurement units**: This is a unit that includes a set of accelerometers and gyroscopes
- **Temperature and humidity sensors**: These compensate for errors due to thermal expansion
- **Angle sensors (encoders)**: These calculate the exact positions of arms and end effectors
- **Load cells**: These avoid system malfunctions and breakdowns
- **Vision systems**: These are used for identifying objects, especially for pick-and-place operations

The following are examples of basic actuators:

- **Stepper motors**: These are used for high accuracy
- **Servo motors**: These are used for better control of the overall system
- **Pneumatic and hydraulic actuators**: These are usually used for higher load capacity

Drives and motors

There are very few machines that do not have moving parts and that do not need drives or motors. Thus, most machines have some form of motion components. There are servos, steppers, variable frequency drives, and motors connected to them for controlling moving parts.

Depending on the precision needed, the machine builder will choose between a stepper, variable frequency, or servo drive. Indeed, the cost implications also change based on the choice of drive and motor.

Robots, too, are controlled by drives and motors. A six-axis robot has six motors and associated drives; a **Selective Compliance Articulated Robot Arm** (**SCARA**) robot (a type of industrial robot) or a delta robot has three motors and associated drives.

All motion components are controlled by the PLC in various forms. There are different ways to control these motion components, and we will take a detailed look at them in *Chapter 5, Motion Control – Control, Synchronization, and the Interpolation of Axes for Accuracy and Precision.*

Programming software and tools

Beyond the hardware components such as I/Os, PLCs, drives, and motors that we have examined till now, there is an important component—software. Software permeates all components of automation. As much as it is a physical presence, the software also determines the function of the parts.

PLC is the brain of the system, but what powers it is the software. A PLC without software is like a car without an engine. PLC is merely a hardware component, and the same piece of hardware is used in different applications. What differentiates the PLC is the software. If for any reason the PLC fails, the hardware can be replaced, and the machine can be brought back to life by simply exchanging the software. This is exactly like when you change your phone; with a few clicks, all data can be brought back to your phone.

As we can see, the software defines the way a system performs and works. Thus, software is becoming an increasingly essential element in machine and factory automation.

A piece of software can be primarily classified as system software, application software, and libraries.

Let us now look at system software. **System software** means the operating system and libraries. Like with every CPU, to address the CPU directly from user code is very dangerous. This is because usually in programming there are variables that are either entered by the user or holding intermediary calculations. However, there is also a possibility to reference the storage space where the variable is stored. With this, we directly use variables to work with or address the CPU from a user code. If such codes are wrongly implemented, there are chances that the CPU will shut off and go into an unknown state. CPUs have functions and code that differ from version to version, revision to revision. The operating system provides uniform access to the CPU. For industrial automation purposes, we need to react to events in real time; hence, we need a **real-time operating system** (**RTOS**). **System libraries** are common functions that are used in every application. The manufacturer generally provides these functions along with the operating system.

Application software is the area to which the automation engineer/programmer devotes the most time and attention. This software is user-specific, and it provides the functionality that the machine builder/machine user actually needs. With the same controller, machines can be made to perform very different functions using different application software. All algorithms that are used for different control purposes are implemented in the application software.

Algorithms and products

Automation is mainly always experienced by automation engineers in terms of algorithms and products. The operating system and other libraries are only of concern to the more experienced programmer. The developer's view of the actual functionality focuses on the logic and algorithms that produce a good product. The definition of a good product (end product), and what features the product must have, is defined by the manufacturer of it, who we will call the end user (from the machine builder's, automation vendor's and programmer's perspective). The end user is usually a manufacturing plant or factory.

From this point of view, automation is viewed by what it actually delivers. It is understood by the functionality. It may be controlling temperature, tension control, or the control of the position of a machining tool tip.

Indeed, the rest of our discussion throughout this book will be about products, which algorithms are used in the manufacture of these products, and how there are surprising commonalities in the algorithms used in manufacturing very different products. At the same time, there are startling differences in algorithms of similar parameters, depending on the way the mechanics are constructed.

You should, by now, have a complete understanding of how control systems are deployed in a machine. Moreover, you should also understand how the smaller elements build up to form the entire system.

Summary

Jacob is taken aback after getting a glimpse of so many aspects he had no clue about and is captivated throughout the conversation. Jacob now has so much more information than he had before. He wonders how, if this information had been available to him before his internship, life might have been a little easier. After this conversation, he now understands how a product gets made, what the role of a factory is, and how machines are deployed in factories to enable production. He also now has a clearer view of various automation elements and how the puzzle fits together and can see the complete picture. However, this is merely an overview, and there are so many aspects and finer details that need to be covered.

This chapter shed light on how we, as human beings, are surrounded by automation while performing all our daily rituals and activities. Moreover, you are now also able to understand the changing manufacturing landscape, how it continues to evolve, and how consumer demands and needs are fueling this change. You were introduced to the elements of a factory, a line, and a machine. Finally, you were taken into the world of automation and introduced to various essential elements, such as hardware and software, comprising PLCs, I/Os, drives, motors, and libraries.

Jacob had a million questions at the start of the conversation, and even after the engaging discussion, he still has millions more. I am sure you, like our young and enthusiastic Jacob, have similar questions in your mind at this point of the book and are looking forward to reading what the father will further expose about the field of automation and manufacturing.

With an understanding of the basics of automation, factories, lines, and machines, we are all set to open a new chapter and look forward to learning so many new and interesting topics.

In the next chapter, we will dwell on specific and interesting control challenges and explain how the manufacturing of everyday products actually overcomes so many control challenges. The first control challenge we take up is in *Chapter 2, The Art of Temperature Control*.

2
The Art of Temperature Control

Global warming is a term on everyone's mind. On a much smaller scale, maintaining the temperature of a material to a specific value can be quite challenging. This is in spite of the fact that temperature is said to be a slow-moving variable. Thus, *managing temperature is an art*. It is one problem to heat or cool the material to the desired temperature. It is a different problem to retain the material at that temperature. This is because natural phenomena, such as conduction, convection, and radiation, continuously transfer heat, which forces temperature change. As temperature changes slowly due to these processes, we must be precise and accurate about calculations. But why do we even fuss about controlling temperature accurately?

In this chapter, we will take you through various manufacturing setups, final products, and assemblies that all need the temperature to be precisely controlled. We will explain how heating and cooling are effectively used in manufacturing products, such as bottles (plastic), plastic parts, metal, fabric, lamination (packaging), and tablet coating (pharma).

The control challenge is maintaining the temperature at the desired value (at a constant temperature). We will explain why such accuracy is needed and how it is achieved in control systems. The chapter will also shed light on how a machine deploys controls, sensors, and algorithms for managing temperature effectively.

We will learn about how temperature can be controlled in the best possible way and how to avoid overshooting, undershooting, and fluctuations.

This chapter will cover the following main topics:

- Highlighting the need for temperature control
- The control challenge
- An overview of applications (plastic and pharma)
- Overcoming the control challenge

Highlighting the need for temperature control

After the fascinating and elaborate initial discussion about machines, lines, and factories, Jacob was already looking forward to his next interaction with his father. During the entire course of the initial discussion, Jacob was so deep in conversation with his father that he lost track of time. He had already started thinking about questions based on his understanding and preparing for the subsequent discussions with his father. Josef didn't need to prepare for these discussions as he was a master of the subject with years of experience. Josef was delighted that he could spark an interest and instill confidence in Jacob, who had lost all interest in studies just a few days ago.

The moment Josef called Jacob the following day after dinner, Jacob sprang from his chair and sat down on the couch where Josef was seated. It was as if he had been waiting for his father to call him to explain new concepts. He had not been so excited lately, and an evening out with friends was no longer pivotal; his discussions with his father had higher priority. Josef was undoubtedly happy with his son's enthusiasm to learn new topics.

Jacob already knew from his previous discussions that his father would now dwell on specific and exciting control challenges with a correlation to the manufacturing of everyday products. He also was aware that the first control challenge to be taken up by his father was controlling temperature. The moment Jacob seated himself on the couch, he was so excited that he started pouring out a string of questions—*Why?*, *How?*, *When?*, and *What?*. I am sure, like Jacob, you too would have many questions on your mind. Before starting the subject, Josef had to calm his son's excitement, and then he could proceed.

The challenge and need for temperature control

Just imagine that you need to boil water at home for preparing tea or coffee, or to drink on a cold, rainy afternoon. Since elementary school, we have all known that the boiling point of water is 100°C. What if you were told to use a thermometer and stop heating when the temperature was 60°C. Sounds easy, right? You monitor the temperature on a thermometer, and the moment it hits 60°C, you switch off the burner below the vessel. The heat source has been turned off but, surprisingly, due to the properties of heating and inertia, the temperature continues to rise by a few degrees more. As the required temperature is 60°C, you add ice to it, and the temperature drops. However, you might notice that the temperature now drops below 60°C. So, you start heating it again. But with previous experience, you stop a few degrees before 60°C. Yet, the temperature shoots up a little. So now you wait for some time until the temperature reaches 60°C. However, you soon realize the temperature stays at 60°C for just a couple of seconds (or more, depending on the size of the vessel) as the environment affects it. The room temperature will affect the water temperature. Similarly, the size of the vessel affects the rate of cooling. Hence, it involves a fair amount of trial and error to simultaneously heat and cool the water so that the water stays at 60°C.

It is tough to maintain water at a specific temperature. Usually, in industry, the temperature must be controlled within a tolerance band. Different applications need different temperatures with different desired tolerances. Engineers face this massive challenge in designing, developing, and programming closed-loop systems to monitor and control the temperature in many applications in many industries. We will explore the actual control challenge in detail in the following section.

Like heating water, monitoring it and controlling its temperature sounds easy but has many variables. For example, suppose the water is held in a cylindrical container. If the temperature of the water is higher than that of the environment, heat will continuously flow from the hot body to the colder environment. If you heat the water directly, say by using an immersion coil heater, then the water will be hotter than the container. So, the container will draw heat from the water and, in turn, become warmer. But the container is surrounded by air that is cooler. So, the air will cool the container and become warm. On the other hand, if we heat the water by heating the container, say by setting it on a stove, then the container gets heated up first and, in turn, has to transfer heat to the water. Even after the stove is switched off, the hot container will continue to transfer heat to the water, thereby increasing its temperature. So, it is clear to see that it is not an easy problem to solve.

Industry has plenty of examples that look easy but are immensely complex. Moreover, we all know that heat cannot be stored and that it transfers easily as well, which makes it an increasingly convoluted process.

Let us take the example of manufacturing plastic **polyethylene terephthalate** (PET) bottles. As discussed in the previous chapter, *Chapter 1*, *Automation Is a Part of Our Daily Lives*, PET bottles are blown into their final form from a small preform. These preforms need to be heated so that air can be blown into them with either high or low pressure to give them the desired form. If the preforms are not heated uniformly to the right temperature, then the process of blowing PET bottles will malfunction. If the preforms are underheated or overheated, then while blowing, the preforms do not take the desired shape, or the blowing pin runs through the preform, making a hole in it. Thus, the blown PET bottle is rendered useless, leading to wastage. Hence, heating plays a vital role in preparing a PET bottle.

Let us look at another example from the products we use in our daily lives. Any plastic item is manufactured using the blow molding or injection molding process. These products could be a cup, vessel, chair, water tank, automobile part, or accessory, among countless other plastic products. An **extruder** is an area inside the machine where plastic granules are heated to form a semi-liquid plastic, which is pushed (injected) into a mold at high pressure to take the form of the mold. If the plastic is heated inaccurately, the molding process fails, leading to wastage. The desired temperature in the extruder barrel is not uniform throughout but rather must have a gradient. This is because the viscosity of the plastic depends on its temperature. Added complexity arises because the temperature setting in one zone affects the temperature in the next zone.

These are just two simple examples that highlight how temperature plays a vital role in manufacturing countless products. At this point, you must now have realized that controlling temperature is not an easy task. There are so many variables that play a huge role and affect the precise control of temperature. So far, you have been taken through the challenges you might face in real life or in industry for accurately controlling temperature. You have also been introduced to some of the applications in industry that need temperature control. I am sure you must be wondering why the process sounds so simple yet unimaginable when it comes to its complexity. Let us dive deeper into control challenges and how automation helps in overcoming these challenges.

The control challenge

We explained in the previous chapter how sensors play a vital role in sensing the actual conditions in the field. Many sensors are available to measure temperatures, such as **resistance temperature detectors** (**RTDs**), thermocouples, and thermistors. Many industrial applications deploy thermocouples and RTDs. RTDs are more precise and accurate. A typical setup can be viewed in the following figure:

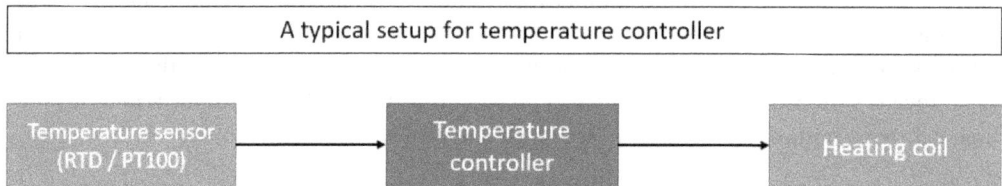

Figure 2.1 – A typical setup to control temperature

If the heating area is greater, there could be several temperature sensors and an array of heating coils:

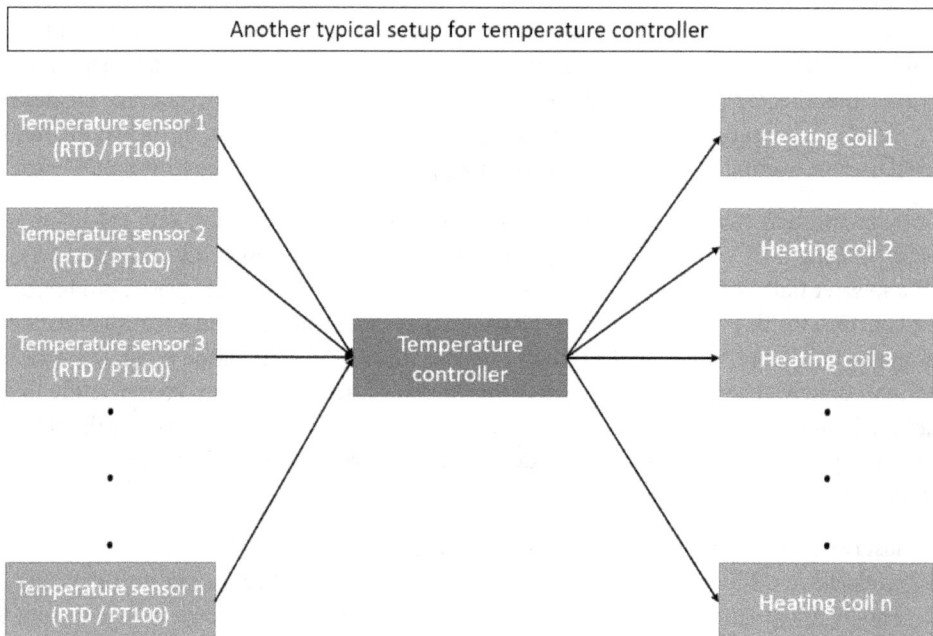

Figure 2.2 – Another typical setup for controlling temperature

The temperature controller could be a dedicated unit or part of the **programmable logic controller (PLC)**. Both configurations are shown in *Figure 2.3* and *Figure 2.4*:

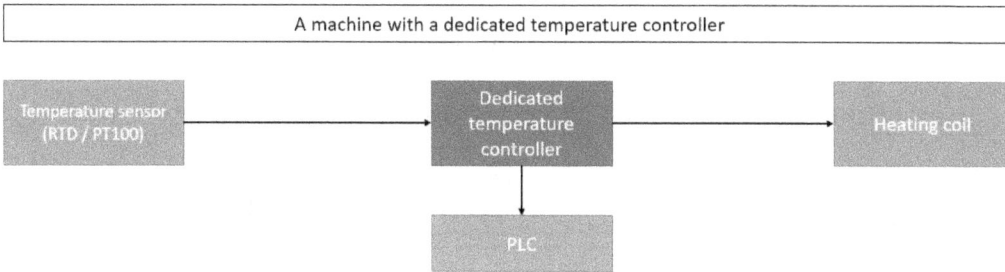

Figure 2.3 – A setup where the PLC and the temperature controller are separate

An integrated controller has greater benefits as the temperature control is integrated into a PLC, providing complete control to the user and software developer. However, low-cost PLCs might not offer an integrated controller where the machine builders need to have a dedicated temperature controller and exchange information with the PLC using additional digital inputs and outputs:

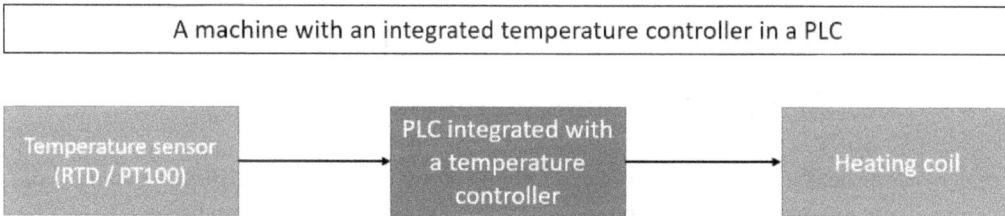

Figure 2.4 – A setup where the PLC and the temperature controller are integrated

These block diagrams provide a very high-level overview of implementing temperature control. However, there are many more elements in the system. A PLC might have onboard **inputs/outputs (I/Os)**, or there might be some distributed I/Os connected over a serial communication bus (RS-485, Modbus, DeviceNet, Profibus), Ethernet communication bus (TCP/IP, Modbus TCP, Ethernet/IP), or a real-time communication bus (Profinet, POWERLINK, EtherCAT, or Sercos), as depicted in *Figure 2.5*:

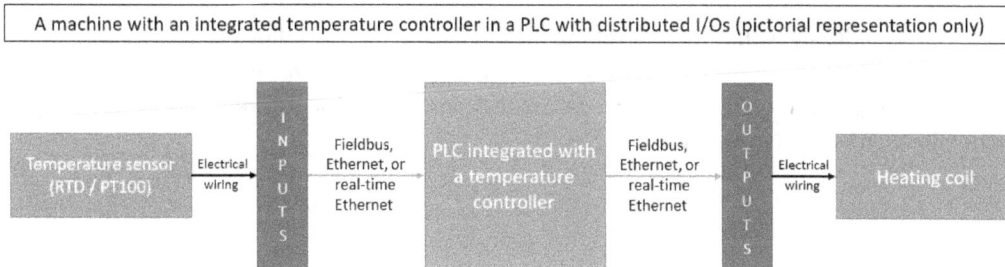

Figure 2.5 – A PLC with distributed I/Os for temperature control

Figure 2.5 is a simplified representation only; in a real machine, it is not how the system would look. This is, however, an easy way to understand how the system with I/Os is placed, making it easy to comprehend.

The following diagram shows what the scenario in a machine would look like:

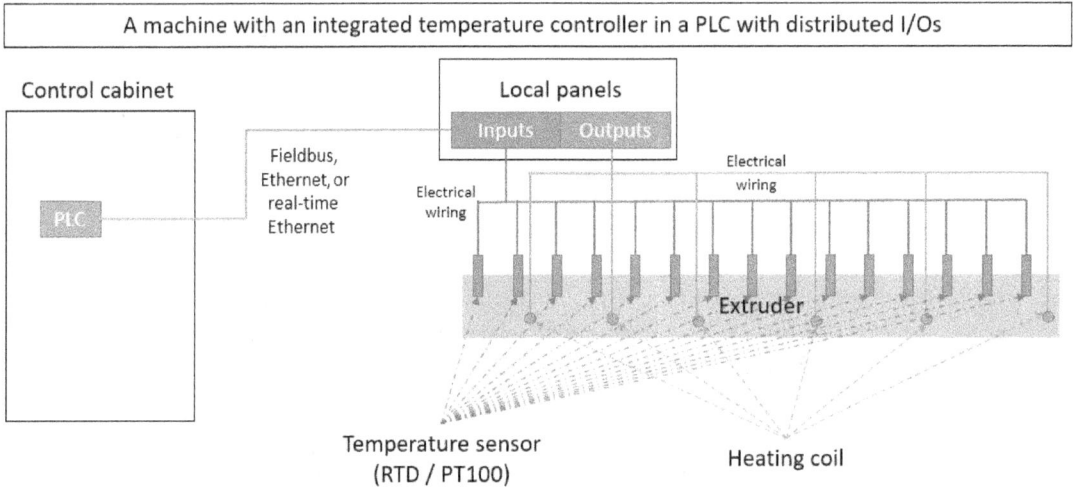

Figure 2.6 – Actual layout of a machine in a factory

A typical machine in a factory can be between 5 feet and around 20 feet long, depending on the type of machine. These numbers are just for you to get an idea of the size; machines can be longer. Usually, a machine has a single control cabinet that houses the PLC. Ideally, most PLCs have an IP20 protection class. Thus, they need to be enclosed inside a control cabinet. The I/O modules can be either inside the same control cabinet, having a centralized architecture, or can be distributed in remote panels. It depends on how the machine needs to be built. If the machine is large, it makes sense to have the I/O modules near the sensors, reducing the cabling and having just one communication cable traced from the control cabinet to the local panels. *Figure 2.6* shows an array of temperature sensors connected to input modules in the local panel and an array of heating coils connected to output modules in the same cabinet.

We will look into the algorithm for controlling the temperature in the following sections and also look at how the PLC handles the temperature accurately and precisely. The temperature is measured by sensors mounted on the extruder connected to the heating coils. The controller goes through the sequence flowchart shown in *Figure 2.7*:

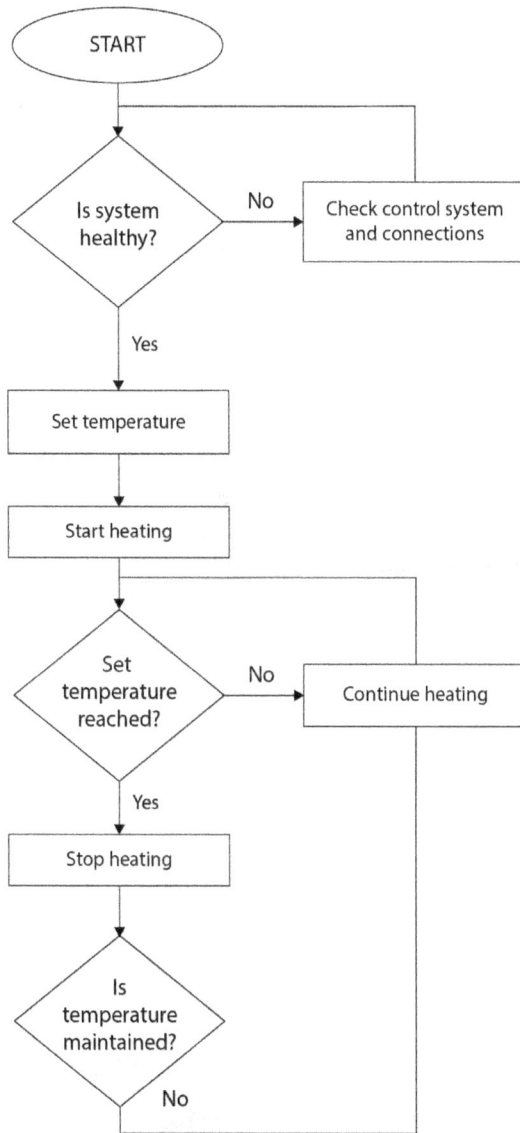

Figure 2.7 – Flowchart for temperature control

The controller needs to handle this flowchart continuously to maintain the temperature based on the closed-loop feedback from the temperature sensors.

In this section, you were introduced to the elements of temperature control and how these elements work together as a unit in a machine and industry. You first understood the control challenge at a very high level, and then you explored the finer elements of a temperature controller in a machine. You were also introduced to the different ways of setting up a temperature controller—a special controller and an integrated temperature controller in a PLC. We finally detailed a flowchart of a typical application and how temperature control works in a machine.

An overview of an application (plastic)

Let us consider an injection molding application for manufacturing plastic components used at home or in automobiles. It could be making plastic cups, chairs, car dashboards, doors, or handles. Many things can be manufactured on an injection molding machine. Depending on the size of the part, the machine and its power increases, and the mold size also increases. Changing molds and manufacturing multiple components on a single machine is also possible.

Let us consider plastic cups being manufactured in an injection molding machine. The plastic granules are put in the hopper, and then these granules are pushed through a cylindrical barrel surrounded by heating elements. These heating elements melt the granules. A force provided by an electric motor or a hydraulic arm pushes the liquified plastics into the mold, which is pressed together so that plastic takes the form of the mold. A coolant cools the molten plastic in the mold, and the final product is manufactured. Just like an injection injects fluid into the body, the injection in the injection molding process injects molten plastic into the mold–hence the name, injection molding machine. You can see a pictorial representation of this in *Figure 2.8*:

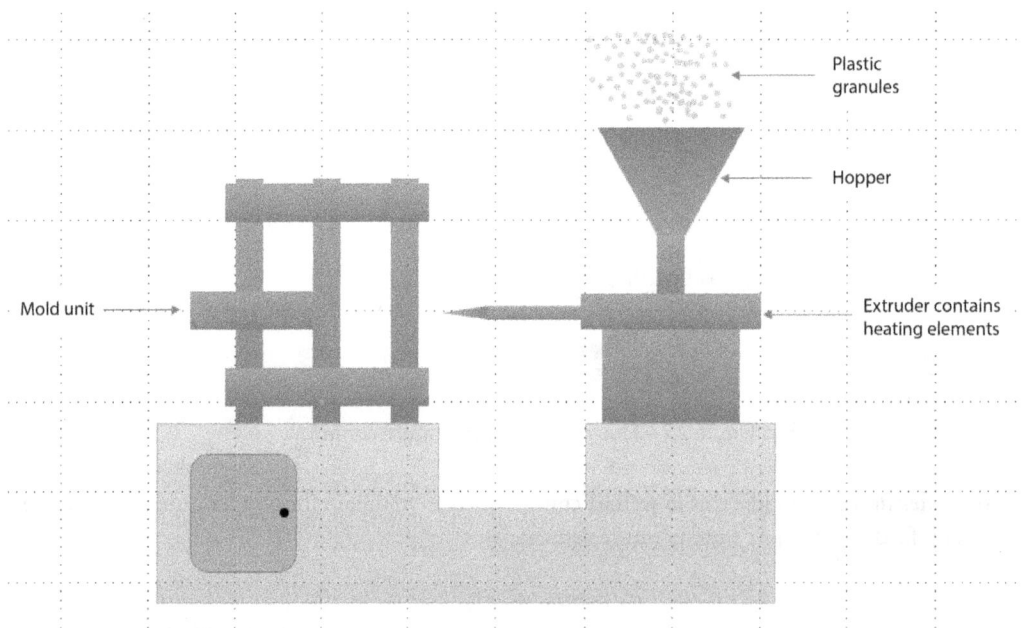

Figure 2.8 – Overview of an injection molding machine

If the heating coils do not function properly, the plastic will not melt, and the injection process cannot be completed. In addition, if the plastic does not melt, it might exert more pressure on the hydraulic arm or the electric motor, tripping it and stopping the process altogether.

The cylindrical barrel has an array of heaters for heating and multiple temperature sensors to provide accurate feedback. The PLC keeps monitoring the actual temperature and, based on the feedback, controls the heating element's supply.

Figure 2.9 shows the steps involved in injection molding:

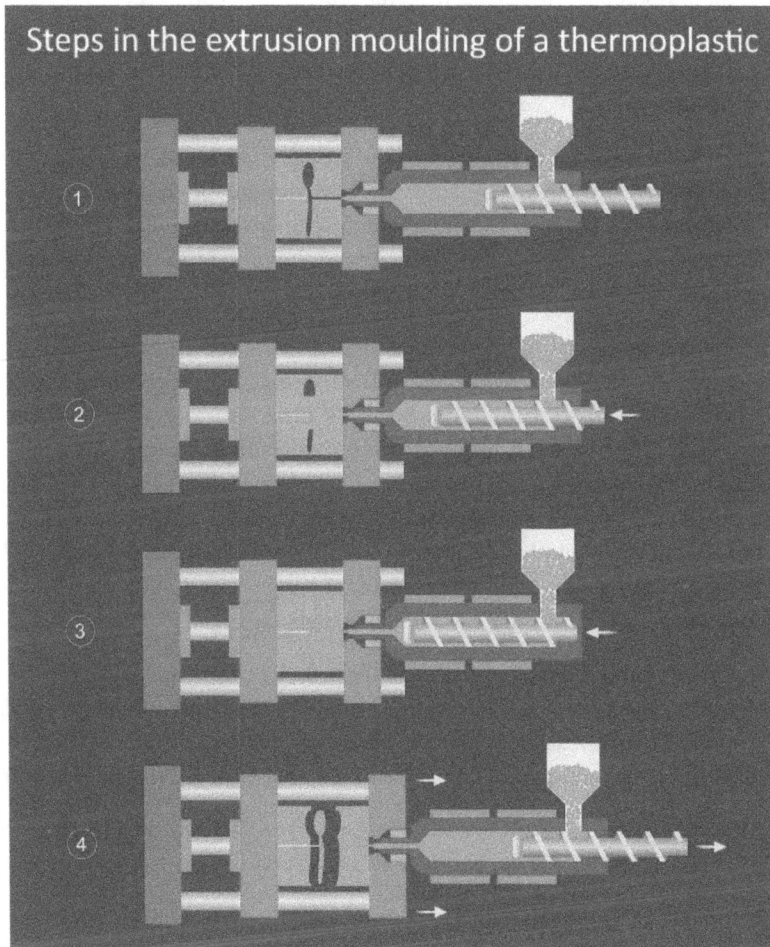

Figure 2.9 – Steps of an injection molding machine while preparing a plastic component

In the following subsection, we will look into the algorithm and how the temperature is controlled in any industrial application.

Overview of another application (pharma)

In many pharmaceutical industries, there are production halls where the room temperature has to be maintained within narrow limits. This is a requirement because the properties of chemical substances are quite temperature sensitive. Some items are volatile and might evaporate when exposed to a higher temperature. So, quite apart from the temperature control of reactor vessels, it is important to maintain the temperature of the production hall itself.

The production hall is usually a large area, perhaps 500 to 2,000 square meters. The room is enclosed, but there is always some amount of air leakage from outside through entry and exit points. The room temperature is controlled by a ventilation system, which is part of a large **heating, ventilation, and air conditioning** (**HVAC**) system. The room temperature is regulated by injecting cold air into the room from the **air conditioner** (**AC**) vents.

There are temperature sensors placed in the room. If the measured temperature is higher than the set temperature, the compressor in the AC is switched on. The compressor compresses the AC coolant and then rapidly evaporates the coolant. This rapid evaporation gives a chilling effect, and this cools the air that is in circulation. The chilled air is pumped into the production hall. More air is sucked in, and the cycle goes on.

When the desired temperature is reached, the compressor is switched off. This discontinues the chilling of air in continuous circulation. After a while, when the room air becomes hotter, the compressor switches on once again, and the process continues. So, this control algorithm is an *on-off* process. One thing that is important to note is that it is necessary to have a **deadband** in the setpoint. That means that if the desired room temperature is, say, 20°C, the room temperature should be allowed to vary between 19-21°C. So, the chilling compressor will switch on when the temperature crosses 21°C in an upward direction, and it will switch off when the room temperature crosses 19°C in a downward direction. This prevents a continuous on-and-off action in the compressor. If the compressor is switched on and off frequently, such as several times a minute, it will get damaged very quickly.

Another interesting observation is the temperature of the product (room air) is not controlled by the heating or cooling of the object itself. It is achieved by **dosing**, that is, by injecting cold air into the room. The temperature of the cold air does not vary; it is only the time duration for which the cold air is injected, and with that, the amount of cold air injected, that is varied to achieve control.

Do note that this description of the *AC* process is merely meant to illustrate the principle of an on-off algorithm. A realistic AC would have more equipment, such as cooling towers and blower fans.

In this section, we explained the control challenge in detail and why temperature control is essential in industry. We explained two major applications, namely an extruder application in the plastic industry and an AC application in the pharmaceutical industry. We also elaborated on the control elements in the temperature control system and differentiated between integrated temperature control and dedicated temperature control. We showcased how temperature control utilizes RTDs and the PT100 sensor and how they are connected to the machine for a typical extruder application. We were able to explain the control challenge and various elements of temperature control deployed in industry.

Overcoming the control challenge

The simplest and cheapest form of controlling temperature is an on/off control. However, this control technique does have certain limitations, with the temperature overshooting and undershooting around the set temperature. This is exactly how we might have controlled temperature during our earlier experiment of heating water. However, the accuracy level is too low, and owing to overshoots and undershoots, there is a chance that the material being heated will be damaged, making it unusable. These fluctuations are unacceptable in many industrial applications. There are also applications where the heaters are assisted by coolers. Similar to the digital outputs connected to heaters, digital outputs are connected to coolers, and the controller takes the action of switching on heaters and coolers according to the need for temperature control.

The following diagram shows the typical functioning of an on/off control. We show the set temperature is 120°C, and the temperature needs to be controlled at 120°C. With an on/off control, we can see that the heater is switched on until the temperature increases and reaches 120°C. The moment it reaches 120°C, the heaters switch off. However, as detailed earlier, the temperature continues to rise, and there is an overshoot. At some point, the temperature starts dipping, and the heaters continue to stay in the off condition. As soon as the temperature drops below 120°C, the heaters switch on. However, it takes some time for the heaters to respond and start the processes of heating, thus leading to an undershoot. As the heaters are now switched on, the temperature starts rising. This process continues. However, the temperature will never be stable at 120°C, and there will always be fluctuations:

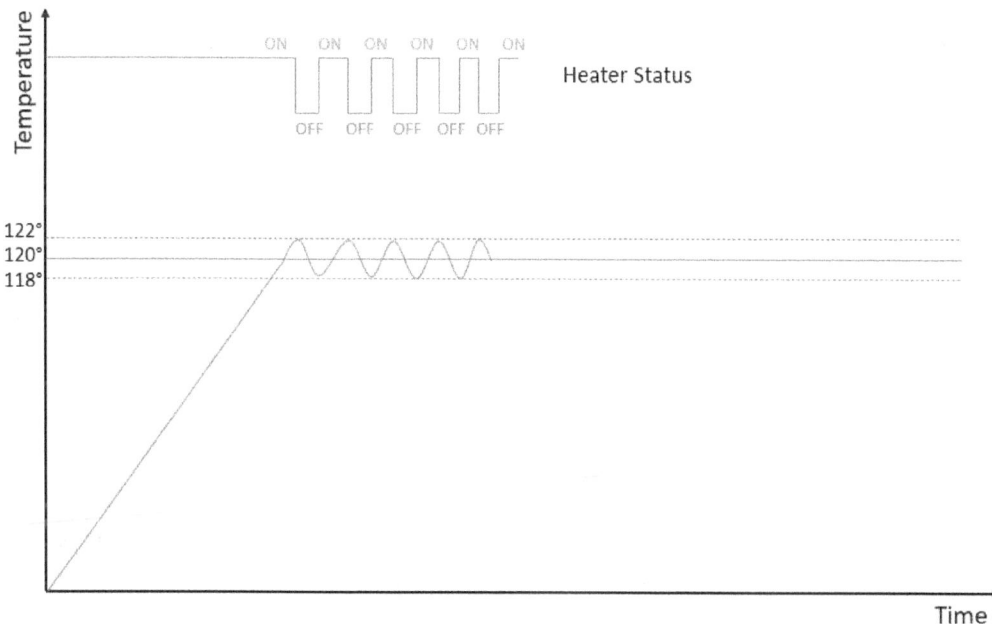

Figure 2.10 – On/off control in a real application and the heater behavior

There is another option that overcomes this issue. A typical industrial application can deploy a **proportional integral derivative (PID)** temperature control for accurately controlling temperature. With the help of a PID controller, it is possible to define how quickly and how much correction should be applied to maintain the temperature using different values of proportional, integral, and derivative gains.

Let us look in brief at how the PID control functions separately so that we get an idea about how it works together. *Figure 2.11* shows how a reference value is set and how an error signal is translated to the reference value:

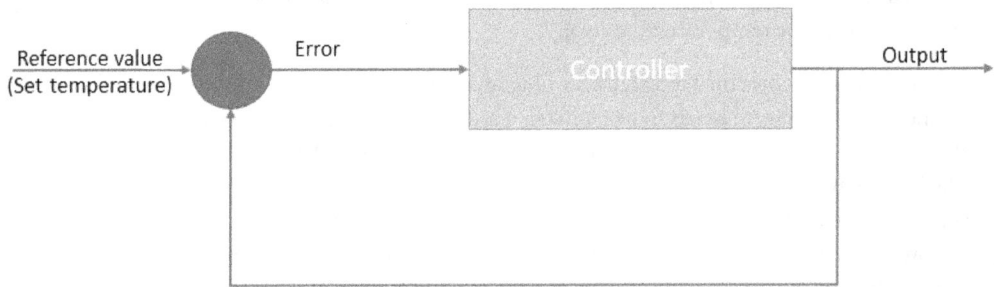

Figure 2.11 – The typical working of a temperature controller with reference and error

This is the basic working of a temperature controller. The key is to design it in such a way that the errors reduce with time. PID controllers are efficient, effective, and easy to deploy in the field. Proportional, integral, and derivative define how the error is treated or handled:

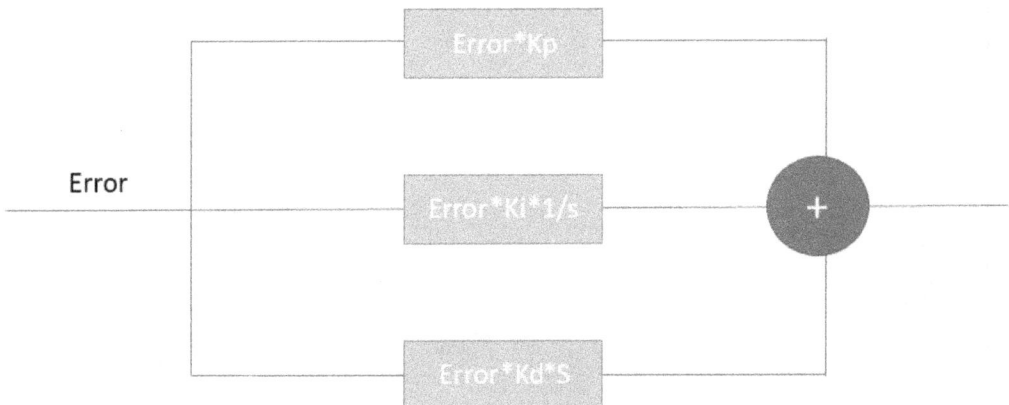

Figure 2.12 – How an error signal is treated by the PID controller

The proportional part in the PID provides a control signal based on the error multiplied by a constant value (**Kp**). In the integral part, the error is continuously summated (integrated) and then multiplied by a constant value (**Ki**), whereas in the derivative part, the error is derivated (which means the rate of change is calculated) and then multiplied by a constant value (**Kd**) to calculate the control signal. The integral component is needed because, usually, a residual error persists when we use pure proportional control. Integral control focuses on this residual error and provides a control component based on the persistent error. The differential component provides correction to react to sudden changes in the error. It works at the rate of change of the error and yields a quick response so that the overshoot or undershoot is minimized. These Kp, Ki, and Kd constants are called **gains**, which can be tuned for efficient working. Thus, by changing the gain values, it is possible to tune the PID to the specific sensitivity needed for an application in a machine. It is also possible to use any combination in the PID controller, that is, if we set any gain to 0, then that element will provide a 0 output. Thus, we can use a PI, PD, ID, P, I, or D controller, depending on the application's needs.

I am sure you have a question in your mind regarding how you find the right values of the Kp, Ki, and Kd gain. Do you need to do the mathematics yourself? Luckily this is not the case. All PLCs have a software library to tune your PID controller based on the system setup. This tuning needs to be done once while starting the system for the first time. The process of tuning is needed to determine the gain parameters that will help the controller provide the desired output. This helps to minimize the fluctuations in the output. Once the tuning is done, the function block in the PLC provides the Kp, Ki, and Kd gain values as output. This can then be connected to the main temperature controller function block in the program. This process of tuning the PLC to identify the parameters of Kp, Ki and Kd is known as auto-tuning.

An auto-tune functionality is provided, but it is possible to set the Kp, Ki, and Kd gain manually, too; however, a good number of calculations are needed, and several tests have to be conducted. It is also possible to auto-tune the controller and then also make minor adjustments manually to make the controller as sensitive as desired. A typical PID function block is shown in *Figure 2.13*:

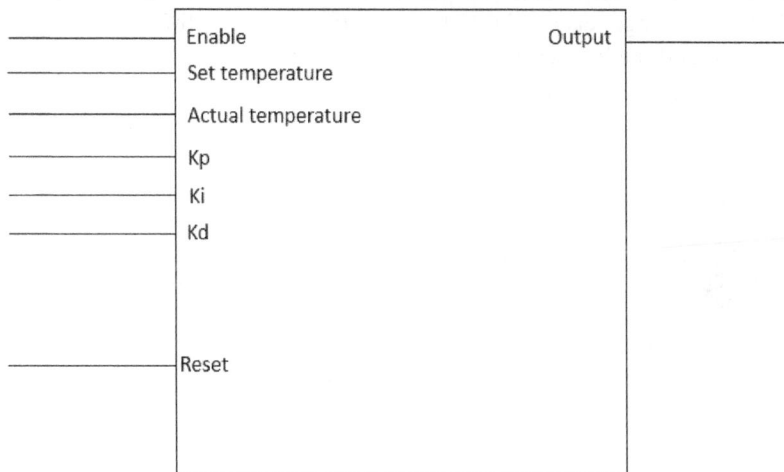

Figure 2.13 – The typical function block of a PID

This is merely a pictorial representation, and depending on the brand of the PLC used, the software function block does undergo changes. However, the representation is to provide you with an idea of how the system functions in a PLC environment.

What would happen if the temperature sensor were faulty? Well, an alarm would need to be generated, and the process should stop. The reason is that without feedback, a closed-loop system cannot function. Thus, if a temperature sensor is faulty, then it would return an abrupt value based on which our PID controller would take action, and as you must have guessed, the values from the calculation would be totally wrong.

These parameters can also be given as inputs on a **human-machine interface** (**HMI**) with a touch screen. With the latest control systems, it is also possible to have connectivity via mobile phones and tablets. A typical setup of the HMI is shown in *Figure 2.14*. The HMI is programmed as per the needs of the user, and the software developer enables the programming of the HMI and the PLC. This helps the operator to easily interface with the PLC. The HMI is also a place where all the errors and alarms are displayed, based on which the operators can take the required actions:

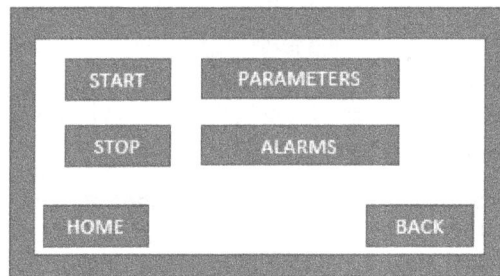

Figure 2.14 – An HMI for human intervention

The HMI is usually a standalone, unintelligent unit connected via a communication cable to the PLC. The communication exchanges data and variables with the PLC. The HMI acts as an I/O device. In some cases, the HMI and the PLC are integrated devices and are provided as a single unit; that is, *the PLC and the HMI are the same*. A typical setup of controls in a machine is shown in *Figure 2.15*:

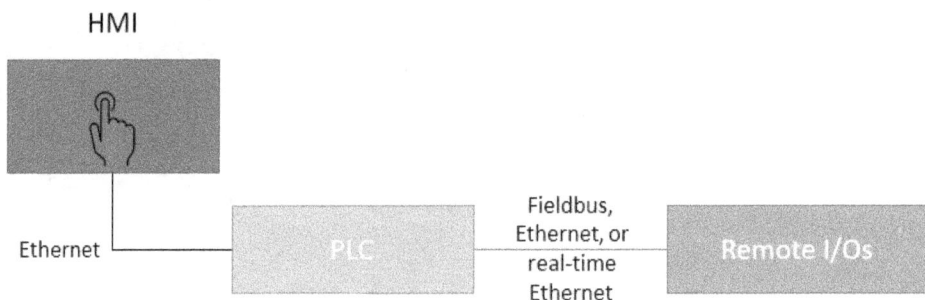

Figure 2.15 – A typical setup of a machine with PLC/HMI and I/Os

These settings can also be hosted on a central server, **supervisory control and data acquisition (SCADA)**, or a **distributed control system (DCS)**, and the parameters can be pushed to the local controllers via a data connectivity mechanism. In addition, there might be **enterprise resource planning (ERP)**, a **manufacturing execution system (MES)**, or any other IT infrastructure that connects with the SCADA or DCS. There is also a possibility that multiple products with different needs for temperature control need different parameters. You can tune the controllers for different needs and have these settings stored in the controller. Depending on the product on the machine, the gain parameters can be loaded for immediate operations.

A machine is just one element in a factory. The machine (PLC, HMI, and I/Os) fits into the larger ecosystem of a factory. The factory has a SCADA/DCS connected to the IT infrastructure. A typical setup of a factory is shown in *Figure 2.16*:

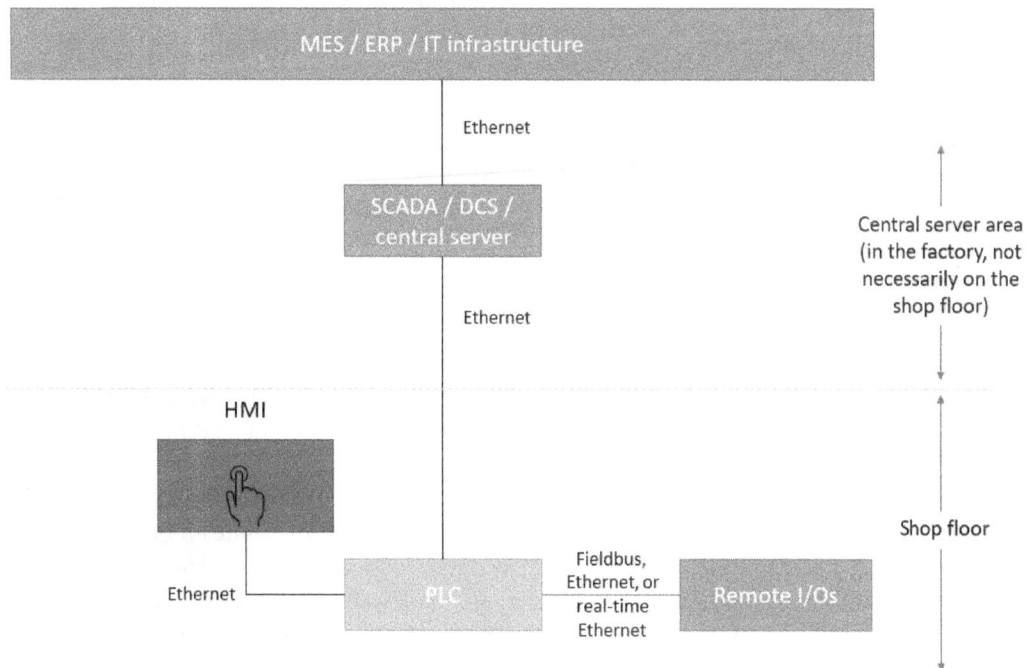

Figure 2.16 – A shop floor with a machine, a central ERP/MES system, and a SCADA or a DCS

After looking at the challenges in controlling temperature, it is necessary to look at the various options to overcome the challenge in industry. In this section, we detailed various applications where temperature control is used and how devices overcome the challenge of controlling temperature. We introduced the concepts of a PID controller and how the PID control helps in regulating the temperature with minimal fluctuations, as compared with the on/off control. We then explained the concepts of programming and auto-tuning and how they are allied to industrial automation. Finally, after elaborating on these concepts, we introduced you to the architecture of a machine and a factory.

Summary

In this chapter, we saw industry examples for temperature control and the challenges faced by machine builders and software developers to build accurate and precise temperature control systems. We initially introduced the challenges any temperature application brings. We introduced this challenge by taking a day-to-day application of boiling water, which set the tone of the chapter. We then navigated and explained how temperature control functions as a control challenge. After you gained a basic understanding of the subject, we moved on to showcasing how temperature control is achieved in actual machines with a controller, either a dedicated or an integrated one. In addition, we also saw how I/O modules play a role in measuring temperature and how heaters are controlled. We also introduced you to an application and how a typical application works in the field or a machine for manufacturing products using a flowchart. Finally, we introduced you to how the PID controller works and how it is programmed on an actual system.

Jacob had never thought that a simple task of heating could be so critical in industry. He instantly remembered the time when, during his internship, his reporting manager was infuriated when he heard that some maintenance personnel had fiddled with the temperature parameters. Jacob remembered that to restore the gain parameters, his manager had to work through the night. He also needed support from the machine builder as well as the automation company that had developed the machine program. Jacob now had an understanding of why these parameters were so important to his reporting manager and the operation of the system. Moreover, the machine on the shop floor was not operational for hours and, therefore, there was no production on the system. This led to delays in production and deliveries as well as generating a lot of waste. Jacob was now able to relate how his textbook knowledge of PID control was applied in industry. Additionally, he was now able to understand not only the challenges in building a temperature control closed-loop system but also what happens if the temperature control goes wrong in the field or if someone changes the parameters.

Jacob was then even more eager to continue the conversation with his father, but as it was getting late, they both hit their beds. Jacob was looking forward to his next conversation with his father. His father informed Jacob that he would next take him through tension control and show him how tension control is effectively deployed on machines for various applications.

3

Tension Control – Managing Material Tension

Tension! Stress! These are some words that we encounter quite often. Primarily, this arises from managing our time or performance expectations during study, work, or any activity at hand. This tension sometimes feels unpleasant, but it is also needed to foster achievement. However, did you know that tension or pressure is also exerted on materials, either on finished goods or in production? Just as we manage stress and tension in our personal lives, materials have to manage tension effectively. In the process of manufacturing several products, tension is an essential parameter. In production, if tension is not managed correctly, it usually leads to material deforming or breaking. Stress beyond a limit on finished goods or products would lead to the product breaking or being damaged. In all such situations, whether it's tension on a human being or tension on a material, it must be kept within strict limits. We will take a closer look at how tension is managed in manufacturing processes.

In this chapter, we will take you through various manufacturing setups, final products, and assemblies that need precise tension control. We will explain how tension on materials is effectively used in manufacturing products such as wires and thread (textiles).

The challenge in terms of control is maintaining tension within the desired value (constant tension). We will explain why such accuracy is needed and how it is achieved in control systems. The chapter will also shed light on how a machine deploys controls, sensors, and algorithms for effectively managing tension.

We will learn how tension can be controlled in the best possible way and avoid material deformation, damage, and breakage. We will cover the following topics in this chapter:

- The need for tension control
- The control challenge (tension control)
- An overview of an application
- Overcoming the control challenge (tension control)

Jacob returned from school, full of tension because of impending exams. However, rather than preparing for his exam, he was wondering if such tension was also a factor in industrial processes. After an exciting discussion about temperature control, Jacob was now looking forward to his interaction on tension control with his father, Josef. After an initial introduction to understanding the details of temperature control, Jacob was curious about how tension control is managed on an industrial scale. He had no clue that tension control would be a critical topic in industry. During his previous interaction with temperature control and a discussion on tension control, he had already started forming questions based on what he had learned and prepared for the subsequent talks with his father. Josef didn't need to prepare for these discussions, as he was a master of the subject with years of experience. Josef was delighted that he could spark interest and instill confidence in Jacob, who had lost all interest in studies just a few days ago.

The need for tension control

Let us use another analogy involving some of our day-to-day activities, which will make you realize that tension control is not too complicated. Just imagine a rubber band or a small piece of thread. If we pull it with a small force, then the rubber band elongates, which is visible to the naked eye, whereas a thread does elongate to a certain extent, although it might not be visible. However, what would happen if we applied more force in opposite directions? The rubber band, as well as the thread, might snap and break, which we do not wish for.

This brings us to the point of applying the right tension to a material and maintaining it so that the material does not break, and a planned process can be carried out on it. Like all control problems, the solution consists of measuring the control variable – in this case, the material tension – and trying to increase or decrease the tension so that it approaches the set point. Thus, tension control plays a vital role in industrial applications, which we will explore in this chapter.

How would you pull an object, tow a car, or even tie smaller things together using strong knots? Usually, we use a rope, and if the object is smaller, we use a thread. I am sure you all must have witnessed that if the object is heavier and the rope cannot withstand the stress, the rope breaks. Another example is tearing a piece of paper, which sounds like a simple task compared to pulling objects or the task in the previous chapter on heating water to a specific temperature. However, the tear is the result of the stress applied to the paper.

What we understand from the prior examples is that anything breaks or tears if the amount of stress or tension subjected to it exceeds the threshold. The stronger the material, the larger the tension or stress needed to break it. This strength of the material is called **tensile strength**. An indication of this is derived by finding the tension at which the material will break or snap. This is called **breaking tension**. In real life, you can imagine tearing a single piece of paper, as previously mentioned, quite easily. However, try tearing five pages – a little challenging; try tearing 50 pages with two fingers – very tough.

What would happen if you held a paper in your hand and tried to pull from both ends? It would most probably tear in two. However, would it be possible to exert force from both sides and pull it apart without tearing it? This sounds challenging. You would need a lot of focus to keep the tension constant without tearing the paper. Industry professionals experience a similar challenge. In a factory or a manufacturing scenario, machines along with production lines transform raw material into finished products. When raw material is transformed into a final product, different machines work on the product at different stages of the production line. An overview of this was provided in *Chapter 1*, *Automation Is a Part of Our Daily Lives*, in the *Inside the factory* section. For the manufacturing of many products, such as newspaper printing, film, packaging, and the winding and unwinding of textiles, it is required that products are maintained at a certain tension.

> **Do bear in mind**
>
> When we talk about tearing paper, we are not talking about tearing in the cross section; rather, we mean grasping two ends and exerting tension along the axis. Cutting along the side of the paper is called shearing, which is a different phenomenon.

Do you know how newspapers are printed? The process of printing a newspaper is tedious and lengthy, involving multiple elements. Let us focus on the aspect of tension or stress. Later in this chapter, in the *Overcoming the control challenge (tension control)* section, we will give a detailed explanation of how a newspaper is printed and the criticalities of newspaper printing. A roll of paper is passed through several rollers, where it is printed with content. You might have envisioned that the paper would have to be maintained at some amount of tension, and this tension needs to be managed accurately. You can easily imagine that printing on paper with some creases on it is not ideal. As well as newspapers, a few more examples where tension control plays a vital role will be elaborated on.

Clothes are products we simply cannot live without. Clothes make us look good and presentable, whether we are getting ready for the day, hitting the gym, going for a run, heading out for work, or dressing up for an evening at the bar.

Let us look into the question of how automation and tension control relate to making clothes, printing, and newspapers. As technical people and engineers, have we ever considered how clothes are made or newspapers are printed? Any process for manufacturing the aforementioned products is lengthy and complex.

Control challenge

A typical setup for monitoring tension control in an industrial environment is elaborated on in the following diagram. It is not too different to what we saw in the temperature control example, but the difference is the way it is measured. The following diagram provides only a high-level overview of an application:

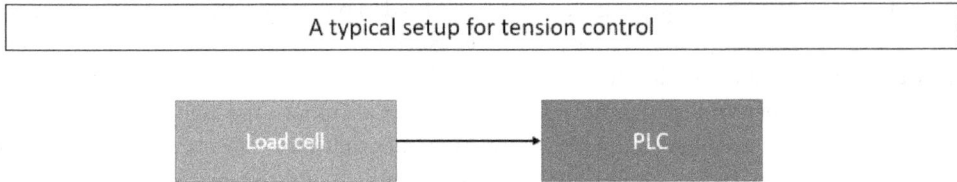

Figure 3.1 – A typical setup for measuring tension control

Meanwhile, a machine in the industry would have the following architecture:

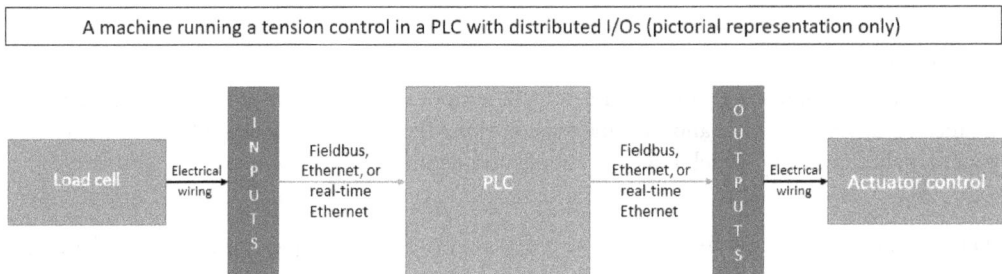

Figure 3.2 – A machine architecture running a tension control application

The preceding example shows the use of a load cell for acquiring field information and sensing the load/stress/tension on a particular product. This is one form of a closed loop and how the sensor inputs can be used by the **Programmable Logic Controller** (**PLC**) to monitor the load on a product, such as a wire or thread. The load cell measures the load on a product on a machine. These values are analog in nature and are fed to the analog inputs, connected to the PLC via cables. Some PLCs have I/Os on board the control unit, and in this case, there is no need for cabling from the I/O unit to the PLC. This was elaborated on in *Chapter 1, Automation Is a Part of Our Daily Lives*. The PLC contains the logic/algorithm that then provides the appropriate commands to the actuators to take certain actions. If there is an error, such as a product break, then the tension suddenly drops; if a product gets stuck in mechanical parts, then the tension suddenly increases. These are criteria for error handling, and a machine is stopped immediately after an alarm is sounded for handling alarms and checking the problem.

As compared to our challenge elaborated upon in the previous chapter on temperature control, tension does not take long to build up and respond to changes. Thus, in this challenge, reaction times are of primary importance. Even a minor delay in responding to changes in tension can result in material wastage, losses, and delays in production. This eventually reduces process efficiency and productivity. In the case of the paper application that we will look into in the upcoming section, *An overview of an application*, even a minor delay in response can lead to paper tearing or a misprint, leading to wastage.

We will primarily look into three solutions for managing, monitoring, and maintaining tension in an application. A minor overview is provided in the following diagram:

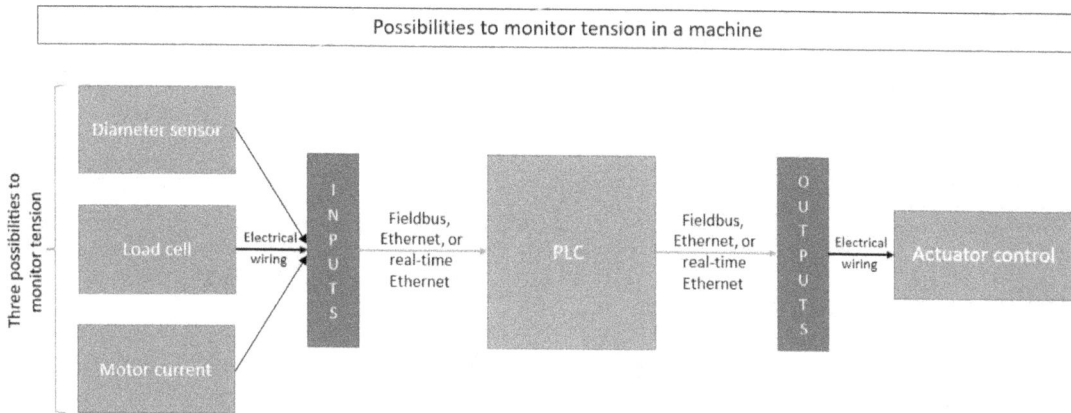

Figure 3.3 – Possibilities to monitor tension

In the preceding diagram, we show how different forms of sensors are used in a machine with input modules. In order to measure tension, the machine builder can either use a load cell, a diameter sensor, or a motor current. All these inputs are received by the PLC, and the necessary logic (algorithm) is implemented so that the correct actions are taken.

In this section, we provided a brief overview of the control challenge. We then introduced you to the use of lead cells to measure the tension of a part or product. In addition to measuring tension using a load cell, we provided an overview of other options to measure tension, using a diameter sensor or a motor current. In the upcoming subsections, we will look at various possibilities to manage tension in a machine with deployed architectures.

An overview of an application

In order to examine a control challenge in detail, we would need to explore an application in any relevant industry. In the following subsections, we will look into some applications, such as textile and newspaper printing applications.

An overview of textile applications

That perfect outfit that you love to wear starts with cotton being made into yarn. This yarn then is woven into a fabric, which then is used to prepare garments. There are so many elements in a production line that are involved in this process, such as spinning, sizing, warping, weaving, printing, dyeing, and inspection, among others. An overview of the elements in the textile production line is provided in *Figure 3.4*:

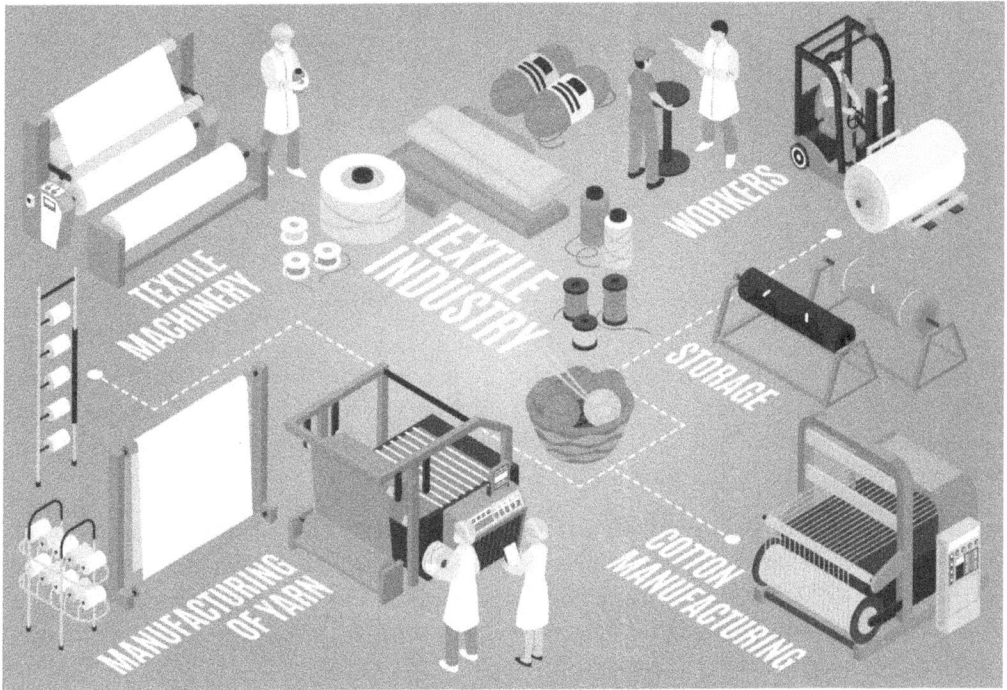

Figure 3.4 – An overview of the various elements in a textile manufacturing line

An overview of the entire flowchart of the textile industry is provided in *Figure 3.5*:

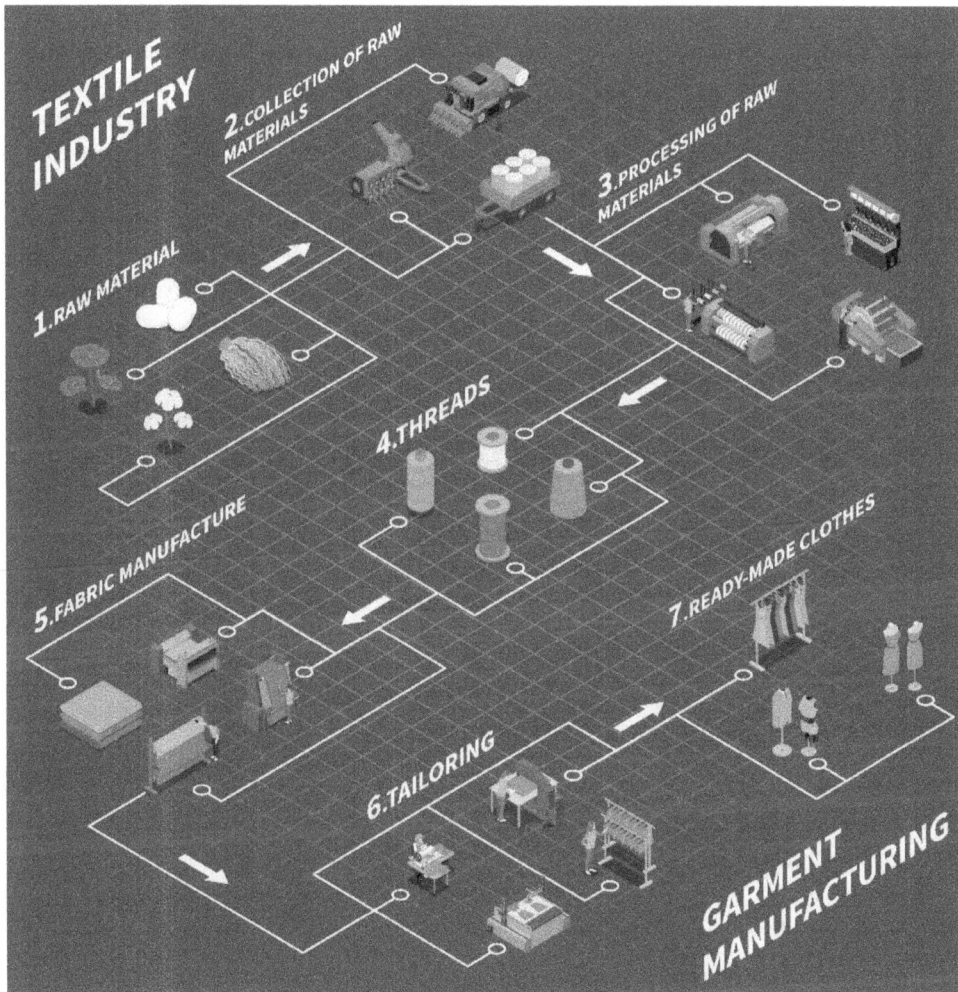

Figure 3.5 – An overview of the flow from raw materials to garment manufacturing

This flowchart explains how the raw material is collected and processed into threads. The threads are used to prepare fabric. This fabric is then tailored, and ready-made clothes are available for consumers to buy from the stores.

Envisioning a fabric-making application

From a raw material such as cotton (although this could apply to silk or any synthetic material), initially, the yarn has to be spun. The machine that does this job is a spinning machine. As an interesting historical tidbit, the spinning machine was one of the first applications of a steam engine, even predating the locomotive. Yarn becomes available on spindles. This yarn, eventually, after undergoing some processes that improve evenness, surface finish, and other qualities such as mercerizing and perhaps dyeing, is wound on huge cylinders. The yarn then gets fed to weaving machines. The control problem here is at both the winding and unwinding stages. The spun yarn must be wound on a drum. This roll needs to have no slack points during winding; otherwise, it will get tangled and knotted up. You might have experienced this if you have played with kites. When your kite sadly gets *cut*, the string hangs loose, and you need to wind it to the bobbin. This winding has to be tight!

Weaving machines need the yarn to be fed at a constant tension. If there is not enough tension (**slack**), the yarn gets tangled. If the tension is too high, the yarn will break. If there is a thread break, the yarn has to be spliced, and this means a loss of time and a production delay. Moreover, a fabric made from spliced yarn is tagged as inferior quality.

Pulling yarn (**unwinding**) from a cylinder poses some non-obvious problems. The tension is maintained by rotating the drum or cylinder at a controlled rate. However, as the diameter of the roll reduces, the linear speed of the yarn reduces. Hence, it is necessary that the speed of rotation has to be continuously varied in relation to the diameter of the roll, in order to provide yarn at a constant speed to the weaving units.

Envisioning a fabric inspection application

A fabric wound on a roll is loaded on an inspection machine. The location of loading the fabric roll is called an unwinder area. The fabric is run through a series of rollers on the machine running through the inspection area and then, finally, wound back on another roll. The inspection area could be either a vision sensor-based automated area or a manual inspection done by an operator (human). The place where the fabric is wound on a roll is called a **winding station**.

During operation, the control system is responsible for unwinding the fabric and, after inspection, winding it back on a roll. This is at the push of a button. The controller is programmed to identify any quality flaws in the manufactured fabric. Once the quality check is complete, the wound fabric is set for transportation to the next workstation or for shipping to a store.

A motor is connected to the unwinding roll, and similarly, a motor is connected to the winding roll. Both these motors need to be synchronized. The synchronization is managed via the drives connected to operate the motors. At the push of a button, the process is initiated. The unwinding station and the winding station start rotating, with the winding station pulling the fabric for winding and the unwinding station pushing the fabric out during the process of unwinding it.

If there are any errors, the machine comes to a standstill, with the operator being alerted about the error. The error could be drive overload, the machine being out of fabric, quality errors, and so on. These can be configured in the controller during programming.

In the textile industry, there are a plethora of winding and unwinding applications, such as when the yarn is being prepared, when the final fabric is being inspected, or the fabric being wound on rolls for shipping. All these applications need precise and accurate tension control to be implemented in automation control systems. If tension control is not precisely implemented, then it might lead to fabric tear, yarn break, or yarn tangle, which results in losses for a factory. If we visualize a winding and unwinding application, then we have a roll that unwinds material (yarn, fabric) onto a machine for various processes. The application could be used to perform either a visual inspection (humans inspecting products) or vision-based inspection (camera and vision sensors being deployed in automation systems for inspecting products) of the material. This would then be wound back on another roll before being moved to a new workstation. Similar winding applications can be witnessed in yarn manufacturing, where produced yarn is wound on small rolls so that they move to the next workstation for processing.

An overview of newspaper printing

For many people, the day does not start without a cup of tea and their favorite newspaper. Today, the printed newspaper is supplemented largely by digital media, and in years to come, the digital medium will be the preferred medium too. However, even today, print copies are still available. Moreover, apart from physical printed newspapers, there are so many other printed materials available on the market. Thus, even though digital mediums are increasingly being used, print media still is in demand. Of course, apart from newspapers, we consume a lot of other print media, such as pamphlets, brochures, magazines, and textbooks. The common application in all these cases is a printing machine. For large-volume, high-speed printing, a web offset printer is commonly used:

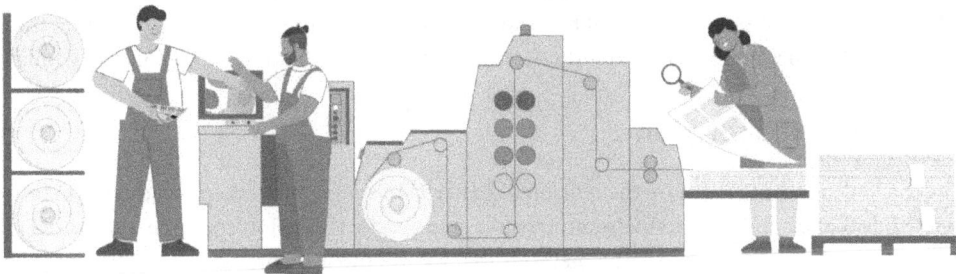

Figure 3.6 – An overview of a printing press

The paper, called a **web**, is mounted on a huge roll. It gets passed through many cylinders to ensure flatness and, you guessed it, proper tension. The newspaper has to be taut so that the printing is registered properly and there are no smudges. What are the challenges in this?

First, as you might have observed, newsprint is not very strong and tears easily. The strength reduces further if the paper is wet. However, the process of printing essentially consists of spraying a mixture of ink and water on the paper, which does make the paper wet. The other challenge is the speed at which the paper travels. Newspapers have to be ready for distribution in the early hours and yet must carry all the last-minute information. So, the time for printing is very short, and for high-speed printing, the speed of the paper processing is high. The paper in a large press might travel at 5 meters per second or more. So, the margin for error is very small – even if the tension is wrong by only a few seconds, many hundreds of newspapers might be defective. Not only is it a waste of paper but also the effort to sort out the misprinted papers before they reach their readers is immense.

For completeness, I will indicate another problem. Right at the start of the batch, the problem of getting the first copies right is quite big. Returning to our example of the kite string roll, at first, the winding on an empty spool is usually not perfect. The same problem occurs later on in a different way. When one spool is about to get empty, printers mount a second roll, which is full at this stage. The paper from the fresh roll gets spliced to the nearly exhausted roll **on the fly** – that is, when the paper is still being printed on. In the transition phase, when the new roll gets superimposed, all load and tension measurements go haywire. Complex algorithms are needed at this point to manage the tension within preset limits.

You can also imagine that at any point in time, if the tension applied is too high, the paper might just tear. This is a major calamity, since splicing and mounting the roll again is time-consuming. It might even lead to a delay in sending the newspaper out for distribution.

Overcoming the control challenge (tension control)

Tension control is indeed critical in many applications and industries. However, there is no one-size-fits-all solution. Thus, monitoring and maintaining tension at a defined level does not have a single straightforward solution. There are many ways in which tension can be monitored in a machine application. The different possibilities will be evaluated in this section. We are not saying that these are the only ways to maintain tension, but these are widely adopted solutions in the industry today.

Ways to measure tension

Let us have a look at how to measure the tension of material in some practical applications. There are many methods that can be used to maintain tension on a product. Tension could be monitored via either *direct means* or *indirect means*. The methods that we will mention provide possibilities and are not the only ways of measuring and maintaining tension but provide a good understanding of how things are managed on machines in the industry. The choice of utilizing one of the methods depends on how a machine needs to be built or the desired response time and accuracy desired by the system.

Direct via loadcell

We elaborated on the load cell approach earlier in *Figure 3.2*. A **load cell** is a kind of transducer where a component such as tension is converted into electrical signals. A strain gauge is a typical sensor used in a load cell. The diagram is replicated here again:

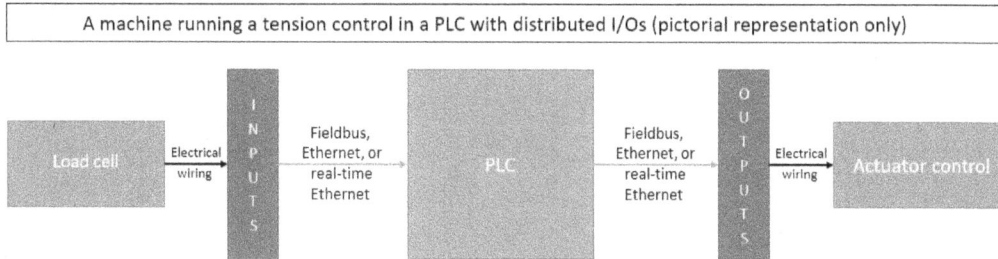

Figure 3.7 – A machine architecture running a tension control application

These transducers are mounted on machines at a suitable area where load/stress/tension can be measured. This is part of machine design and mechanics. These are then wired to special input modules or analog modules using electric wires. However, this is just a way to measure tension. The data received from the transducers is a raw value. The logic, algorithm, or program in the PLC converts this data into actionable information and provides necessary actions to the outputs.

Once tension is measured, the PLC receives some raw data, which needs to be translated into human-readable form. Usually, all transducers provide an operation range; let us take 0 kg to 5 kg as an example. A programmer would translate it as 0 for an input value of 0 kg and 65,535 as 5 kg. Depending on the numbers, the programmer scales the input received and generates a formula. Whenever the input reads a specific value – say, 40,818 – the formula will convert it to a specific value in kg, which would be 3.11 kgs. Thus, the accuracy of the system can be calculated accordingly. This is a typical approach for measuring tension on a product.

However, what if the product does not allow the possible installation of such load cells? How can tension be measured when the process is continuous? Would it mean that it's not possible to measure tension? This is definitely not the case. There are other ways to measure tension.

Indirect via diameter sensor

Another approach to measuring tension is using a diameter sensor. You may be wondering how this is even possible:

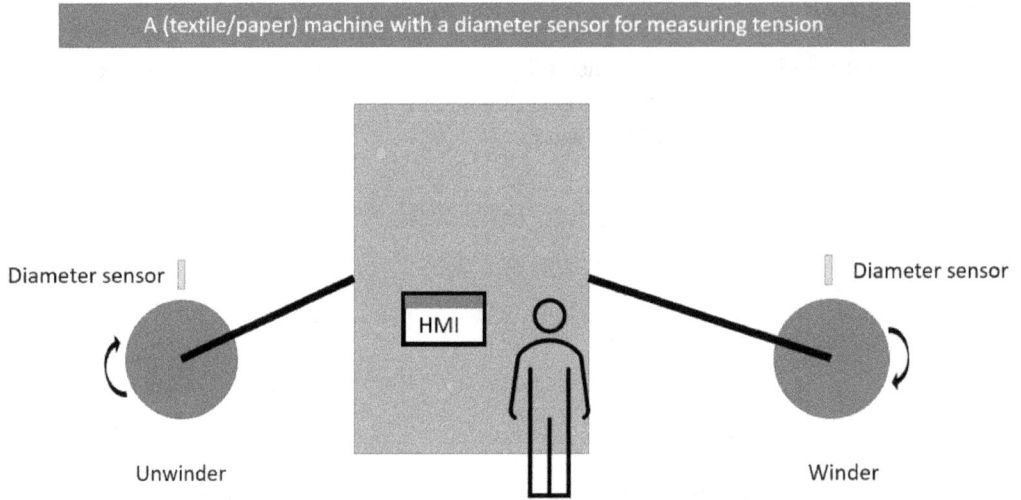

Figure 3.8 – A typical machine setup with a diameter sensor

We have shown a typical setup in a textile machine or a paper machine where a diameter sensor is installed to measure tension. You might wonder why this is even needed. Just imagine you have a paper roll in one hand and are trying to pull and wind it on another roller. As the paper roll is full at one end and empty at the other, if you simply start winding the empty roller, you would need relatively less force, and as the roller starts winding, the force needed starts building up. However, in this case, it is way too easy, as you are winding and exerting pressure only on one side and there is no process in between the rollers.

Let us add some complexity to the scenario. What would happen if you were with your friend and trying to wind and unwind? You are responsible for winding the paper on the empty roller and your friend is responsible for unwinding the full roll. If you are too fast, then the tension on the paper increases, and if you are too slow in winding in comparison to your friend who unwinds, then the paper in between the roll sags.

In this industry, usually, there are not just two rollers but multiple rollers, and there is a continuous process between the systems. This further adds complexity. If there are several rollers and the paper/cloth has to undergo multiple turns, it could be possible that the tension starts building up during this process. This can be seen if you are trying to pull a rope that is stuck between multiple objects; the force needed to pull the rope would definitely be higher than pulling a rope on an open field.

Moreover, in industry, the winder as well as the unwinder have motors and drives for winding and unwinding. It is similar to the analogy of two humans – one trying to unwind a huge paper roll at one end and another trying to wind it, as explained in *Figure 3.9*:

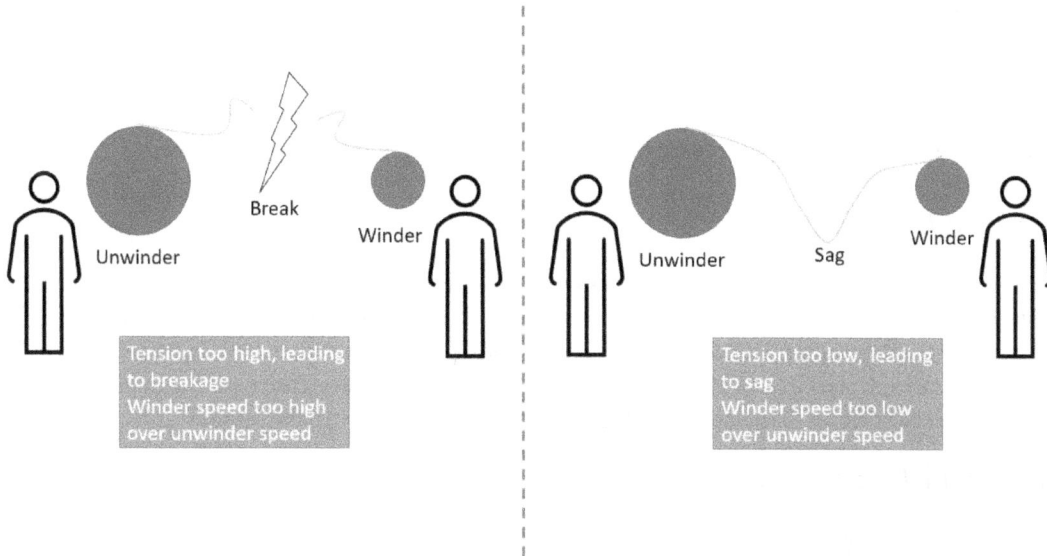

Figure 3.9 – The speed difference in winding and unwinding leads to issues

I am sure you would now have a question about how the diameter sensor plays a role in this situation. The diameter sensor provides information to the PLC as to what the current diameter of the roll is, either on the winding side or the unwinding side. This provides a closed loop for the PLC to increase or decrease the roller speed so that the tension is maintained on the product – in this case, a paper or a cloth. The closed loop from the diameter sensor enables this possibility.

At this point in time, you might also ask, could we measure such tension without any additional devices and sensors? You would be on the right track; there is a possibility to measure tension without the need for such sensors, but it needs a different approach.

Indirect via a motor current

We can measure tension based on the current drawn by the motor, either winding or unwinding the rolls. A motor is driven by a drive, and this makes it possible to read the current drawn by a motor. At normal operation, without load, the motor will draw a certain amount of current to rotate. When it is on full load, the current drawn is higher, and as the load reduces or increases, the current increases or decreases accordingly. At no load, the motor draws minimum current, and at full load, the current drawn is higher. If the product gets stuck, then the motor experiences a higher load, which leads to the motor tripping and stopping the entire process.

This method of measuring tension is also deployed in many machines where load cells and diameter sensors cannot be mounted. The change in current can be monitored faster and the reaction times are considerably slower to changes in tension.

Ideally, there would be ready-made software blocks or libraries to monitor such tension control applications. However, it is also possible to build software on our own using basic coding, such as a ladder, C, or other textual programming methods. Tension control can be implemented using basic code and might not need the utilization of function blocks or libraries. Scaling, in the case of the load cell, is the easiest form of implementing tension control in a PLC.

Varying the tension

As we are aware by now, measuring the control variable – in this case, tension – lets us know what the value of the variable is now. Then, we compare it with the **set point** – the desired value of tension. The difference is the **delta** or **deviation**. If the delta is outside a present band, then we have to undertake corrective action. Corrective action is to vary tension – increase or decrease with the intention to bring tension back closer to the desired value. How shall we go about varying tension?

Control the length of the catenary

The portion of the wire, rope, fabric, or web between the two supported points is the **catenary**. One end of the catenary comes off the feed roller. The other end is the entry to the processing machine itself. The movement process is achieved by two synchronized motors, one unwinding motor to turn the feed roll and another motor to pull the material into the machine. There will be a varying amount of slack due to momentary differences in speed between the two motors. So, the web or yarn is passed over a dancing roller that can move up or down, depending on the amount of slack. The position of the dancing roller gives a signal to vary the torque delivered by the unwinder motor.

Introducing friction by a tensioning device

Another method to increase or decrease tension is by applying friction to the material under processing. Typically, this is managed by passing the material between a pair of rollers. If the measurement of tension indicates a need to increase tension, the two rollers are moved closer to each other. Thereby, the material gets squeezed or pinched, which results in an increase in tension. If the control asks for a reduction of tension, the rollers are moved slightly apart.

In this section, we introduced different ways of managing tension and compared industrial applications, with an analogy of industrial automation. While looking into ways of managing tension, we explored three major concepts and methods deployed in the industry to manage tension. Depending on the application, product, and needs of the user, a particular method can be deployed. We saw that all these methods have their own share of positives and negatives. It depends on the budget available for the machine builder versus the accuracy and precision desired when choosing a deployment.

Summary

In this chapter, we saw industry examples for tension control and the challenges faced by machine builders and software developers to build accurate and precise tension control systems with high reaction times. We introduced you to the various challenges faced in our daily lives, with examples that explained the concept of tension and the need to monitor and manage it well. We then introduced you to the real-life industry challenges and practical scenarios where tension plays a vital role, which if not monitored or maintained leads to waste and losses. We then moved on to provide various perspectives on industry applications, such as paper and textiles. All these topics highlighted in detail the need for tension control and the results if not managed properly.

Jacob was surprised by these insights on tension control. He had never imagined that tension control could play such a big role in the products being used daily. Thanks to Josef, Jacob now had a better understanding of tension control. Jacob wondered how paper splices on the fly and how a machine manages to maintain tension without material tearing. He was impressed with the technology being put to great use in the form of automation. Jacob, in fact, wanted to try out all these things on his own and was thinking of getting himself a microcontroller kit to get hands-on experience with these subjects. Josef was thoroughly impressed by his son wanting to try out these experiments in college.

Once again, Jacob was now even more eager to continue the conversation with his father, but it would have been too much information in a single day. Jacob was looking forward to the next conversation with his father. Josef informed Jacob that he would now take him through level control and how it is effectively deployed on machines and processes for various applications. Jacob was now looking forward to the next conversation that would discuss *Level Control: Controlling the Level of Liquid to Avoid Drying up or Spilling Over*.

Level Control – Controlling the Level of Liquid to Avoid Drying Up or Spilling Over

Are you trying to fill a bottle or a glass with water? The main aspect you would focus on is to fill the bottle or the glass with its desired contents of water, and second, to avoid over-filling—that is, to avoid spill-over while filling. You need to manually ensure that the liquid in the bottle or the glass does not spill over via visual inspection. In addition, what if someone needs a refill? You would need to ensure that as soon as the glass becomes empty, it is refilled without spilling over. Well, all these activities are manual and occur at a relatively low speed. Similar requirements exist in industry as well, and this brings us to the important control challenge of level control, which is important in many industries.

Controlling the fluid level in a tank is an essential element in various industrial applications. There is a parallel in our day-to-day lives with the overhead water tanks that supply water to our homes. If the water level is not maintained, then there are chances of the tank running dry or spilling over. The control challenge is to manage the level of filled fluid (in a tank or bottle) within set limits. The measurement variable is the level of the fluid, measured using sensors, after which appropriate action can be taken by filling or draining as may be required.

In this chapter, we will investigate various applications where level control is required and how industry experts manage this control task. We will look at possible solutions, such as measuring using capacitive or ultrasonic sensors. We will explain how we can ensure accuracy in filling applications and how it is achieved via control systems. The chapter will also shed light on how a machine deploys controls, sensors, and algorithms to effectively manage fluid levels. We will cover the following topics:

- The need for level control
- Control challenge of level control

- Overview of applications
- Overcoming the control challenge of level control

The need for level control

Jacob was in college and was sitting under a tree staring at the college building. He noticed the overhead water tank on the college building was spilling over. The college staff noticed this, and the building supervisor rushed to the terrace and attended to the issue. He was instantly drawn into Josef's shoes and started considering how, this being water, it would be easy to manage and there was little risk involved in touching the fluid. However, what would happen in an industry where the fluid was not water but maybe some corrosive or hazardous fluid? During his previous conversation, Josef had mentioned that his next discussion and control challenge would be on the management and control of fluid levels.

That day, Jacob returned from school with plenty of questions. After the detailed discussions on tension control and temperature control, Jacob was now looking forward to receiving in-depth information from Josef about level control. He was able now to envision how a machine fits into a factory environment and could picture manufacturing setups. Similar to the previous challenges, he was expecting that it would involve some kind of sensors, actuators, PLCs, and control algorithms.

As Josef returned from work, Jacob was sitting expectantly on the couch. Josef realized this and, without heading to his room to freshen up, he put down his bag, made himself a pot of coffee, and sat next to him on the couch.

Jacob narrated the incident of the tank overflow at college to Josef. Jacob finally asked what the scenario in industry would be and how factories manage the levels of fluid in tanks. How do factories manage spill-over of hazardous fluids, or even avoid spill-over of any fluid altogether?

Josef asked Jacob what the capacity of his motorbike was. You might also relate to this example from day-to-day life. Jacob instantly responded with 13.3 liters. He was confused as to why Josef was asking this question. Josef asked whether he knew what the current level of fuel in his motorbike was. He was unsure so Josef asked him to go check. Jacob ran to the parking lot and returned in 2 minutes, and before sitting down, he told his father that his vehicle should have around 5-6 liters of fuel at that point. Jacob was still puzzled as to why Josef was focusing on fuel in his motorbike. Josef now asked Jacob to put 8 liters of fuel in his bike. Jacob was confused and instantly responded that this would be impossible. The fuel pump would cut off before 8 liters could go into the motorbike or the fuel would spill over, leading to wastage. The sensor on the fuel nozzle usually stops the fuel from spilling over. If the sensor on the fuel nozzle fails, then fuel spills over.

Challenge and the need for level control

The challenge of fluid level control is to maintain the level of fluid in a container at a given value. This involves control in the form of two different actions—maintaining the rate of inflow and the rate of outflow. This can also mean filling a container with an accurate volume of fluid. Depending on the viscosity, the fluid levels need to be managed and controlled accordingly. We will explore, in this chapter, the ways by which fluid levels can be managed in industry.

In the automotive industry, on the engine assembly line, the engines need to be filled with oil. There is a machine with a tank of oil that needs to be put into the engine. The machine needs to check the level of oil in its tank and then the level of oil already in the engine. If the primary oil tank in the machine is empty, the machine needs to shift to a secondary tank and raise an alert for replenishment of the primary tank.

The procedure in the paint shop in an automotive factory is similar, and in the printing industry, the inks used for printing are stored in tanks and their levels also need to be constantly monitored. If not, it might lead to poor-quality printing as the inks run out and, therefore, a lot of wastage. As seen in the previous chapter, *Tension Control – Managing Material Tension*, newspaper printing is very fast and a small delay in detecting errors will lead to huge losses and wastage.

In process applications such as blending, where fluids are mixed from different tanks, each tank needs to have sensors to measure the level. If the level is misrecorded, then there might be errors in the process that could lead to the rejection of the entire batch. Another example can be found in packaging in the **food and beverage (F&B)** industry. In a filling line, a given substance is stored in tanks that are used to fill bottles. The source tanks' levels need to be monitored and the containers need to be filled with an exact amount of the fluid. In the pharmaceutical industry, the level of medicines and fluids that are packaged or filled in bottles needs to be monitored. Here, sometimes, the goods that are to be packaged could be solids in the form of powder. In any of the preceding examples, if the level is not monitored properly, it usually leads to rejections for quality purposes, and losses and delays in production.

Just imagine, you have guests at home, and you want to serve them water from your water filter. If the filter is dry, then even after opening the faucet, you will not be able to fill your glasses with water. Thus, you will not be able to serve water to your guests. The issue here is that you did not monitor the level of water in your water filter, and it went dry. Similarly, if the water filter is being filled by opening an inlet faucet, and you fail to pay attention, the water filter will overflow. This is exactly what happens in industry too. In any of the prior applications, if the level is not monitored properly, the tanks either run dry or overflow. In addition, it is impossible for humans to manually check the levels of solid, closed tanks. Automation and the control system behind this process play this role.

In this section, you learned about the basic need for managing, monitoring, and controlling levels, which plays an important role in machines and factories. You were briefly introduced to the need for monitoring and managing levels of fluids in tanks. In subsequent sections, we will look at this control challenge in detail and explore some typical industrial applications along with the workflow.

Control challenge

Two typical setups of the tanks described previously are shown in *Figure 4.1*:

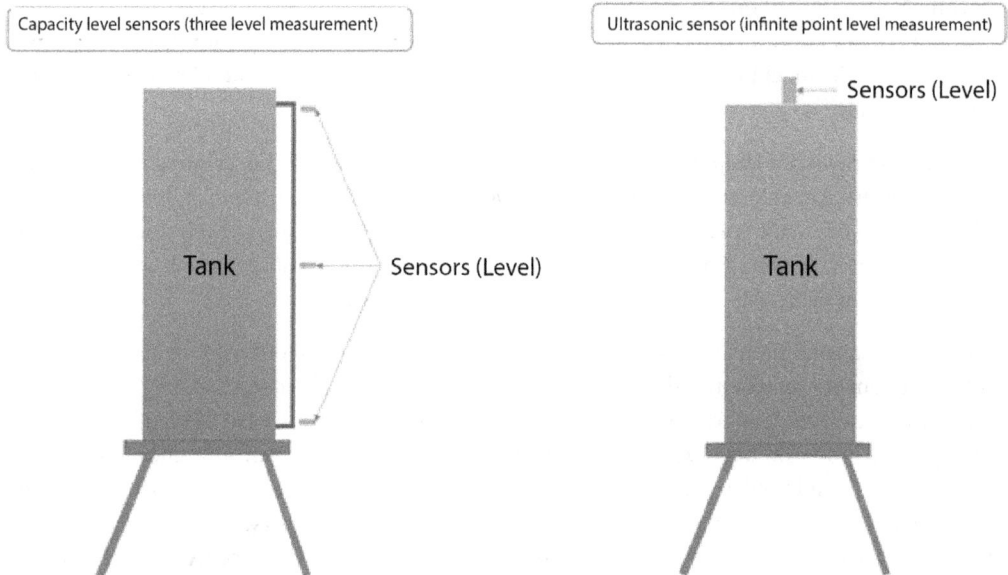

Figure 4.1 – Tanks with level sensors for measuring the level of fluid

In industry, the levels in tanks are measured using sensors. In the preceding figure, we show two possible setups to measure levels. In the image on the left, the tank has two outlets connected with a transparent pipe. Sensors are placed near the pipe at three different levels—top, bottom, and middle. Depending on the application and the factory's need to monitor the level precisely, additional sensors could be connected to the pipe—near the top (90%), at 75% and 25%, and near the bottom (10%). The measuring accuracy itself is high but the programmer, user, and operator get visibility of only three or four levels, depending on the number of sensors selected. If the operator misses the warning at 25%, then the next level about which they will be alerted is 10%, which is very close to drying up. Moreover, if refilling is not done automatically, then there will be delays in production. Let us look at an entire application with automatic filling and cut-off based on the level detected:

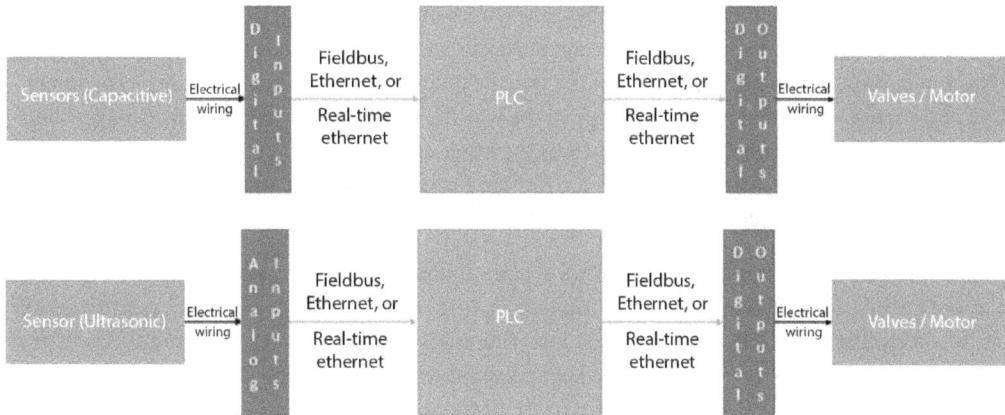

Figure 4.2 – A typical application with sensors and actuators on a PLC

There are two ways of connecting sensors to PLCs: either via digital or analog inputs. If a machine is built using three-level capacitive sensors, then these capacitive sensors are connected to digital inputs. Special modifications are made to the tank so that these sensors can sense the level of the fluid. The user requirements will determine how many sensors will be added to the tank. As the number of sensors increases, the number of input modules also increases, and so does the complexity of the mounting arrangement. All these elements increase the cost of automation devices. A capacitive sensor approach is adopted to reduce costs and yet still provide decent precision and accuracy. One capacitive sensor would need one channel of input for processing the data. Thus, four sensors would require four input channels. Generally, in automation, input modules have anywhere from 4 to 12 channels. Thus, in a worst-case scenario, an entire module might be occupied by only level measurement sensors. Typically, we have seen the deployment of anywhere between two to four level sensors. Just having two sensors for high and low levels might prove inadequate, depending on the rate of filling. If the low-level alert is triggered, by the time the control action of filling takes effect, the tank might already have run dry. On the other end, if the high-level alert is triggered and engages the control action to turn off the inflow of water, the material in the pipeline might still cause an overflow. Thus, if the usage per cycle on a machine is higher than the filling rate, the operation can't be sustained and will lead to errors and production stops. A workaround for such systems is to mount the bottom sensor not at the extreme bottom but a bit higher, around 25%, which gives the user flexibility in utilizing the machine even after the fluid level passes the low sensor.

This is just one way of monitoring and measuring levels in tanks. Another way is to use an analog ultrasonic sensor. We will explain both these methods in detail in the upcoming *Overcoming the control challenge of level control* section.

In either of the previous highlighted approaches, the choice of selecting a particular approach depends on factors such as the cost of implementation and the accuracy needed for the application. Additionally, the choice of approach selected will also affect the way the programmer manages software development. A hardware change will also demand a change in the way the application programmer manages their code, as one approach uses digital inputs whereas the other approach uses analog signals. In general, the tank on the machine is usually connected to a filling system driven by a motor, a valve, or both, depending on the filling mechanism.

The most important challenge in a level control application is to avoid the tank drying up or spilling over as both scenarios will lead to production delays, time wasted, reduced efficiency, and energy losses. In overspilling, in addition to product loss, there is immense wastage of material. There are also applications where the level needs to be held constant within narrow limits by alternately filling or draining the fluid. This is the case in industrial boilers and steam generators.

In this section, we saw how level control is a challenge for machine builders and programmers. We explored a typical tank-filling application with either capacitive sensors or ultrasonic sensors. In the upcoming section, we will have a detailed look into applications of this in industry.

Overview of applications

As explained earlier in the chapter, there are multiple applications where level control is needed and, if not monitored properly, it leads to spillage, wastage, or drying up. In this section, we will look into specific applications in industry that involve level monitoring and control.

The first application is from the automotive industry, concerning the filling of engines with engine oil.

Overview of the automotive application

In the automotive industry, on the engine assembly line, there is a station for filling engines with engine oil as they are being assembled on a conveyor:

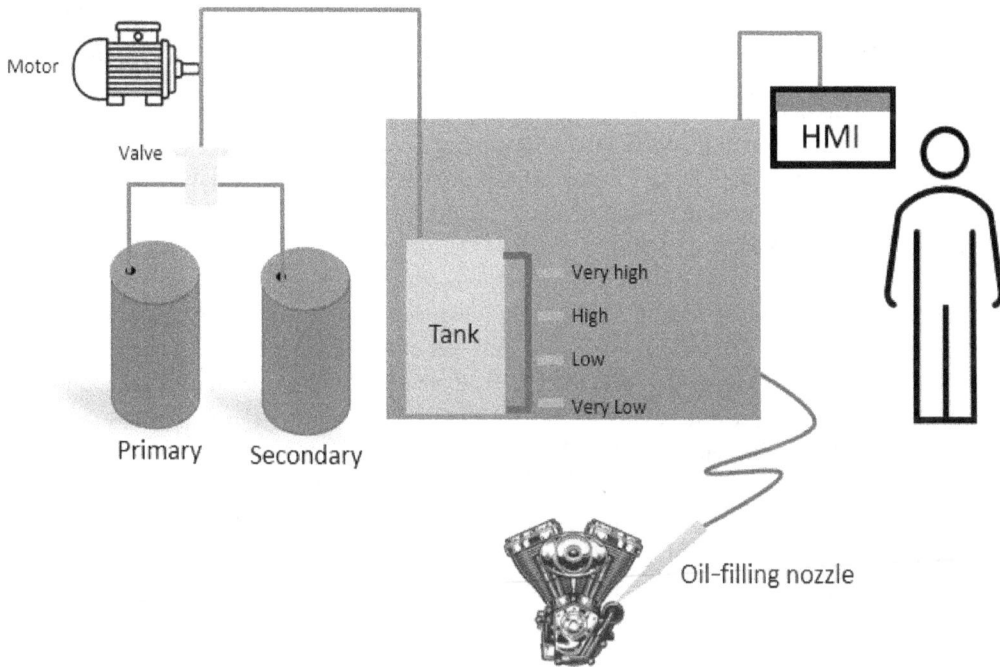

Figure 4.3 – An oil-filling application of level measurement in the automotive industry

The primary objective of the machine is to fill the engine with the desired quantity of engine oil. The volume of oil required is in the order of liters and is monitored using a flow meter. The engine moves from one station to the next on the process line and each station is responsible for performing one operation. Oil filling is among the final stages of engine assembly, after which the engine moves into testing.

Level monitoring and control are applied here. The internal structure on the control side remains essentially the same. There are input and output modules that sense the level of oil in the tank and provide information to the PLC on whether to start pumping oil into the tank. There are two redundant tanks, and the valve connected to the primary and secondary tanks determines which external tank is selected to supply the oil for filling. When one tank is empty, the other tank is automatically selected using the valve control until the empty one is replaced.

When the engine arrives at the station, the nozzle is connected to the engine and the filling process starts. The process usually requires 4-5 liters of oil and takes around 30 seconds. There might be **Radio Frequency IDentification (RFID)** tags on the pallet on which the engine is mounted that store all the data about the engine assembly operations. These RFID tags are checked at the end to verify the quality of the engine. If there are errors, the engine is sent back for rework and corrections.

Let us focus on the level control application. As we see, the primary goal of the machine is oil filling and the tank level control is a peripheral yet essential function. Thus, the control logic is implemented such that if the tank in the machine is full, then all the level sensors (very low, low, high, and very high) show detections of the level (that is, they provide binary 1 as input). Thus, the PLC continues the oil-filling operation and there is no operation in these peripheral systems. In such oil-filling systems, the primary operation is to fill oil in the engine whereas the tank changeover or level measurement act as peripheral systems. Thus, when the tanks and primary and secondary barrels are full, the PLC needs to only perform the engine oil-filling operation and the peripheral systems, such as the level measurement and primary and secondary barrel changeover operations, are at standstill. As oil filling continues, the level in the tank of oil starts dropping. At this point of time, let us assume that both primary and secondary barrels are full. As the oil level in the tank crosses and triggers the low sensor, along with the high and very high sensors (already triggered), this indicates that the tank in the machine is almost empty and needs to be refilled. This signals the PLC to start the motor to fill the tank with oil. As both primary and secondary barrels are full, the PLC first selects the valve position such that as the motor starts, the oil from the primary barrel will be used to fill the tank. This ensures the actual engine-oil-filling application remains unaffected. If the primary barrel is empty, the sensor on the barrel provides information to the PLC, and the PLC automatically shifts the valve to select the secondary barrel. If both the primary and secondary barrels dry up, an alarm should be generated to notify the operator to change the barrels. Thus, in all these cases, level control plays a vital role. If both the barrels dry up and there is no indication and proper interlocks in place in the PLC, the motor might begin operation and run dry, leading to overheating and damaging itself. An **interlock** is a condition in the program in the PLC, which inhibits a particluar operation.Thus, in the oil-filling example, an interlock could be to not start the motor before a particular set of conditions are met. These conditions in programming would be a valve position check, barrel levels check, and tank levels check. Only if all these conditions are met would the motor start operation. If the valve control fails to operate, there are chances of the pipes bursting. All these interlocks are essential for efficient and robust operations of the machine on the shop floor.

In this section, we provided a brief overview of an automotive assembly line with oil-filling machines. You have understood and roughly visualized how an engine is filled with oil during manufacturing. The motive for providing this overview was to highlight an applied use of level control in a machine. You have seen how level control is not the primary function of this machine, but if the level control functionality fails, it leads to line stoppage or wastage. Let us look at how this control challenge is overcome in the upcoming section.

Overcoming the control challenge of level control

There are no complex libraries needed to monitor, maintain, and control levels as we saw with temperature control and tension control. Level control applications are straightforward, and in a way, very elementary. The algorithms are very simple and can be built by new engineers. A programmer does not need to have immense experience to build a level control application. It is essential to focus on interlocks to avoid errors in software development.

Let us look into a typical tank-filling application in industry and see how the systems work as a whole in the machine and what would happen if things went wrong.

In a level control application, we saw in the previous section that the tank could have two to four capacitive sensors or a single ultrasonic sensor. Let us take an example of a tank connected to four capacitive sensors in this section.

Let us assume that the tank has four sensors—full (very high - 90%), high (75%), low (25%), and very low (empty - 10%). The filling system connected to the tank has a valve and a motor. The user needs the motor to run and the valve to open if the fluid level drops to low (25%), and the motor to stop and the valve to close if the level reaches very high (90%).

Thus, in the program, when all the sensors detect a level (input value is binary 1), the motor and the valves are switched off and the machine operates as usual. When three sensors (high, low, and very low) detect a level (binary 1) but the very high sensor does not (binary 0), the motor is still switched off and the valve is closed. When the fluid level falls further, with the two very high and high sensors not detecting a level (binary 0) and two lower sensors detecting it (binary 1), the motor and valve continue to be in an off state. As the fluid level further reduces, the three upper sensors don't detect any level (binary 0). At this point in time, the motor and the valve are activated by the PLC and tank filling starts and the fluid level starts rising:

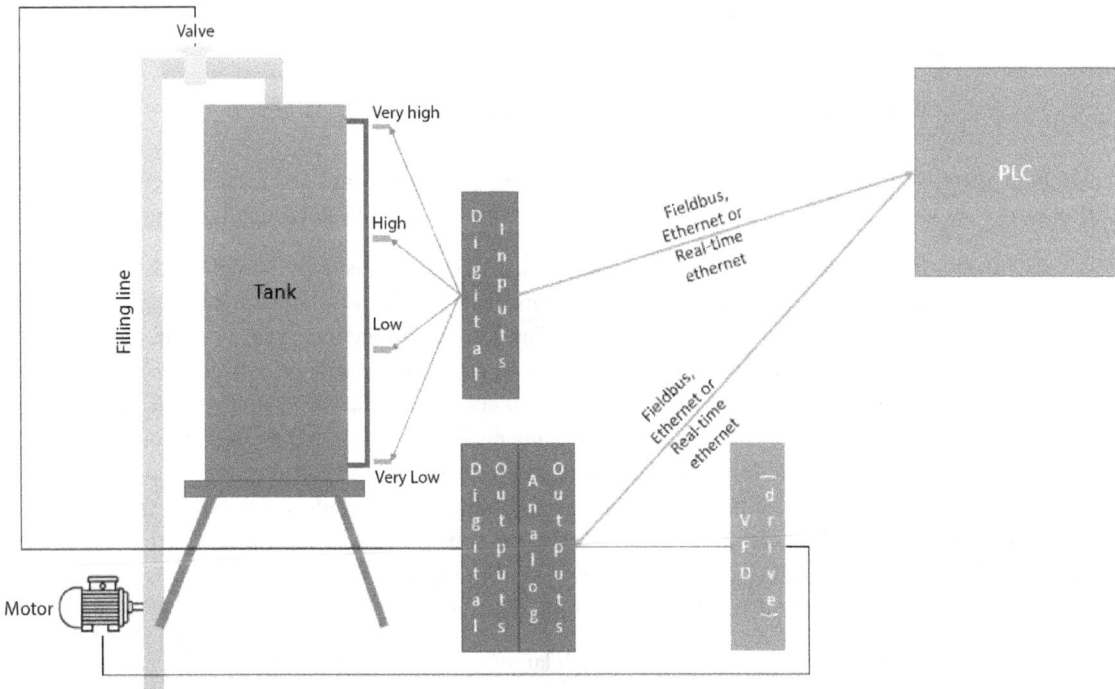

Figure 4.4 – An application with a motor, a valve in the filling line, and a four-sensor setup

This can also be explained in a tabular format. **Digital 1** (binary 1) is for *switched on* and **Digital 0** (binary 0) is for *switched off*:

Tank level	Sensor status	Input status	Motor status	Valve status	Filling status
Very High (90%)	1 (ON)	1 (ON)	0 (OFF)	0 (OFF)	0 (OFF)
High (75%)	1 (ON)	1 (ON)			
Low (25%)	1 (ON)	1 (ON)			
Very low (empty/10%)	1 (ON)	1 (ON)			
Very High (90%)	0 (OFF)	0 (OFF)	1 (ON)	1 (ON)	1 (ON)
High (75%)	0 (OFF)	0 (OFF)			
Low (25%)	0 (OFF)	0 (OFF)			
Very low (empty/10%)	0 (OFF)	1 (ON)			
Sensor malfunction state >> Results in machine alarm / warning					
Very High (90%)	1 (ON)	1 (ON)	0 (OFF)	0 (OFF)	0 (OFF)
High (75%)	1 (ON)	1 (ON)			
Low (25%)	1 (ON)	1 (ON)			
Very low (empty/10%)	0 (OFF)	0 (OFF)			

Table 4.1 – Activation of liquid filling based on the status of digital inputs

Machine operation and tank filling take place simultaneously. Once the tank level reaches very high, that is, all four sensors detect a level, the motor and the valve are shut off, stopping the filling.

It is essential for the programmer to consider sensor failures and disconnections. It might happen that the very low sensor doesn't detect a level (due to some mechanical failure) whereas the other sensors above it do. This shows that the sensor at the bottom has failed to provide binary 0 as input, but the tank is full as all the other sensors show detection. In such cases, the programmer must ensure that the filling does not start. If the programmer has not considered such scenarios, then the filling might start, leading to overflow and waste. In addition, alarms need to be generated to alert the operator about the state of the tank. In addition, if the valve remains closed and the motor starts, the pressure in the pipe from the motor could increase. In this scenario, the motor needs to trip before the pipe explodes. All these scenarios need to be considered by the software developer. The software can be built in different ways depending on the requirements of the user. It is essential that the software developer considers all interlocks before programming the system.

The preceding application is only one part of a much larger system. A tank-filling application would just be a peripheral application. The primary operations are totally different, such as engine oil filling, bottle filling, and so on.

The situation changes a little with the use of an ultrasonic sensor. As the ultrasonic sensor provides inputs in the form of analog signals, motor and valve control can be much more dynamic than step changes of levels. The user can define the exact level of the tank at which it should be refilled. The software developer could scale the sensor to the size of the tank and then accordingly, these values can be set. In addition, using an ultrasonic sensor gives a higher degree of flexibility as the user can decide dynamically when they would like the tank to be refilled.

Let us take an example. The following figure shows how ultrasonic sensors reduce the wiring required as well as adding flexibility for the user. If the ultrasonic sensor provides values of 0 – 65535 for a tank with a capacity of 100 liters, then the value 0 corresponds to 100 liters and 65535 corresponds to 0 liters. Likewise, when the tank is empty (0 liters), the value provided by the ultrasonic sensor would be 65535. Additionally, some sensors allow the programmer to change these values for empty and full. Based on these values, the programmer can scale the values inside the PLC. In addition, the user can set the value on the HMI for when refilling should be initiated. The remaining operations are the same. If the user wants the refilling operation to begin when the tank has 8 liters of fluid left, then when the ultrasonic sensor sends the value corresponding to 8 liters, the filling system is activated, starting the motor and opening the valve. When the tank is detected as being full, the filling stops.

Another difference between ultrasonic and capacitive sensors is in the cost: an ultrasonic sensor with its analog inputs costs more, while the cost of capacitive sensors and digital inputs is much lower:

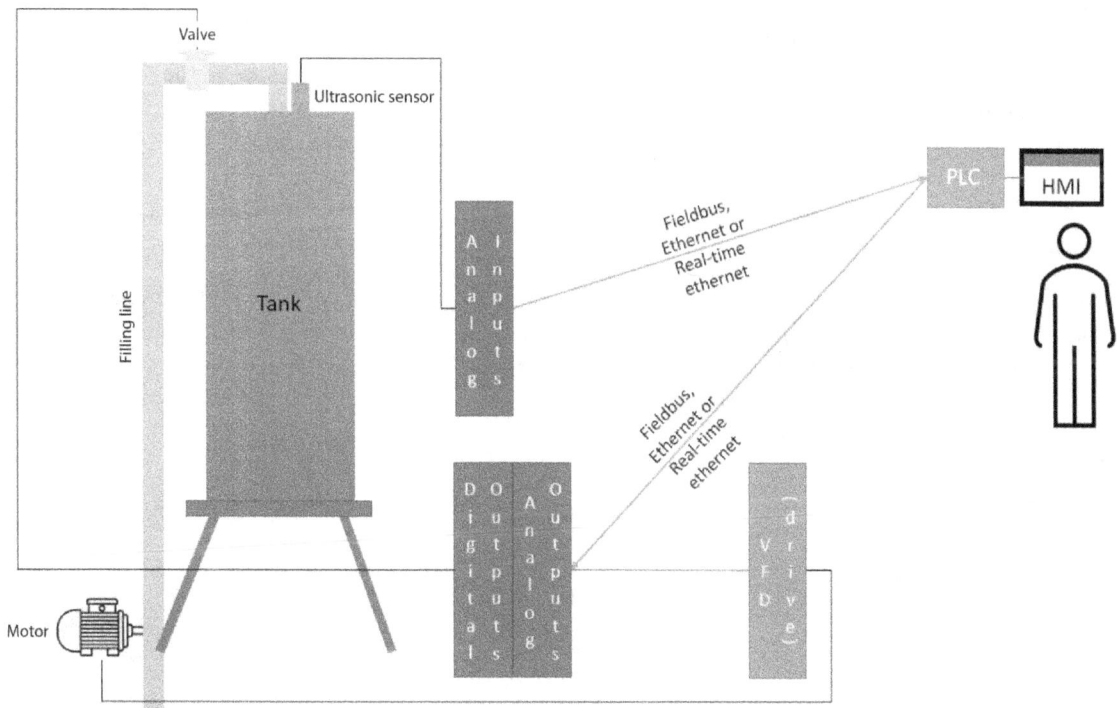

Figure 4.5 – Tank filling application using an ultrasonic sensor

We examined two ways in this section for overcoming the level control challenge, showcasing the use of capacitive sensors and ultrasonic sensors. We detailed how a relevant machine would be constructed and how the sensors are placed for detecting the level. You should now be able to independently visualize and successfully implement a level control application in the industry or at home.

Overview of a boiler application

An industrial boiler is used mainly to generate steam, which is used to provide energy or force in many industries. You could think of a power station where steam is used to turn a turbine. Steam is also useful in the pulp and paper industry. However, we want to look at how these industrial boilers are kept running.

Water is supplied to a boiler through pipelines and is measured as it flows in. Inside the boiler too, there are level gauges that continuously indicate the volume of water. Water is heated typically by burning a fossil fuel, either coal or diesel, or sometimes gas. This converts water to steam at a particular pressure, which is then drawn off for use in an industrial process. As the water gets converted to steam, the water level will drop and needs to be replenished by the flow of inlet water. An industrial boiler is normally a shell and tube arrangement—the water to be heated flows through a convoluted tube inside the shell. The shell holds the heat from the fuel, usually in the form of hot air or flue gases. So water moves in from one direction and gets converted to steam as it travels through the shell. Here, the level control is actually a flow control. It is important that the outlet produces pure steam and not a steam-water mix. If the flow rate of water is too high, it might not get fully converted to steam. It is important in this case to measure the humidity of steam at the outlet, called the **wetness of the steam**. If the wetness is too high, it can lead to corrosion downstream, so the input flow rate needs to be reduced or the fuel to the boiler needs to be increased. The downstream process might need a specific amount of steam. So, we see a complex interplay: the amount of steam needed, the wetness factor of steam, the amount of water being pumped in, and the supply of fuel. This is called a **multi-variable control loop**.

Here we are talking of three measurands, namely, feed water inflow, steam outflow, and drum level. In the control scheme, there are two PID loops—one to control the drum level, while the output is connected to the second PID that controls the feedwater inflow. The aim of this cascaded loop is to minimize the fluctuation of steam outflow:

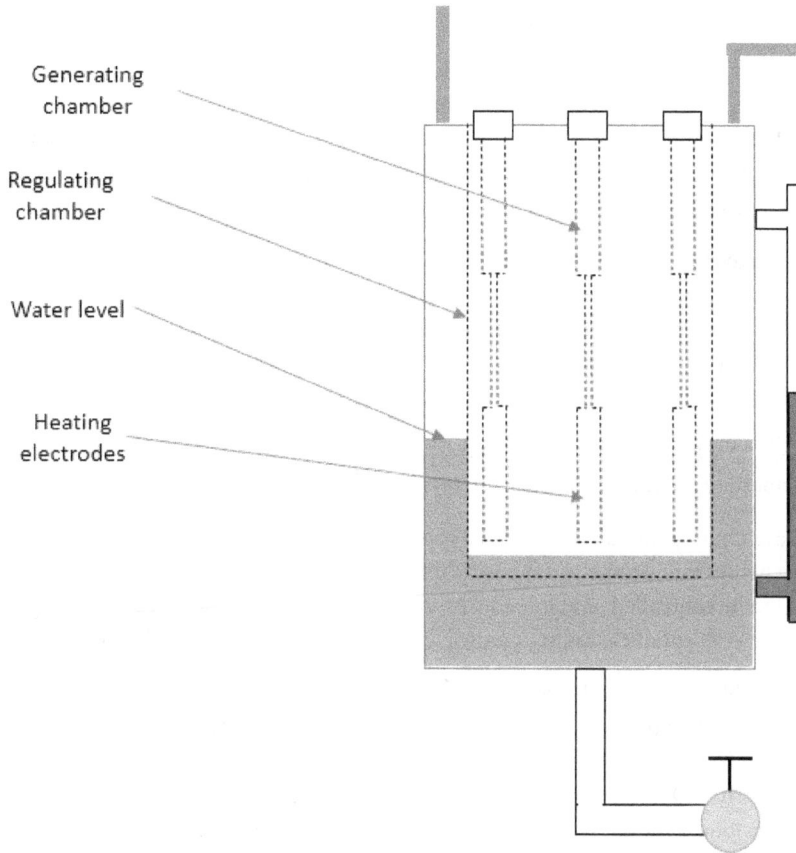

Figure 4.6 – Overview of a boiler setup

In this section, we took you through various applications used to overcome the level control challenge in industry. We provided an overview of various possible solutions for the automotive industry as well as some process applications of an industrial boiler. You have understood how the level is accurately measured and how automation enables the machine to function properly without stoppage or wastage. We also showcased how sensors, either capacitive or ultrasonic, are deployed to provide inputs on levels and how the PLC responds to such inputs.

Summary

In this chapter, we saw some industry use cases for level control and the unique challenges faced by machine builders and software developers to maintain and monitor levels in industrial applications. We saw that, in such applications, reaction times are not too critical. In addition, we saw that level control applications are not the primary function of the machines involved, yet if the level control applications fail or do not perform as desired, it can result in machine stoppages or even wastage.

We introduced you to simple level control applications such as the overhead tank filling that we see in our daily lives, which gave you a perspective of the challenge. We then provided an analogy of industrial applications and detailed the concept of level control and ways of maintaining and monitoring it. We also took you through the results if level control is not managed properly. We then elaborated on an application in the automotive industry involving engine oil filling.

We gave a brief glimpse into a more complex application of a boiler control using cascaded loops—the classic three-element drum level control.

With that, we have demonstrated an important aspect of industrial automation. We can always have simple control schemes to monitor and measure one parameter of a process. But in real life, multiple parameters need to be controlled simultaneously. Variation in one parameter causes changes in other parameters. So, the concept of cascading control and automation schemes are necessary.

Jacob was instantly able to connect the dots and relate to the overflowing overhead tank. He understood that there must have been a malfunction leading to the spill-over. He also realized the adverse effects of such an overflow in industry. Just like you, Jacob was thinking of the effects of such an overspill in a fuel factory or the chemical industry. He was delighted that Josef had taken him on this level control application journey and explained to him so easily how level control applications work and how the critical challenge is overcome. Thanks to Josef, Jacob was now ready for any level control tasks that he might face in university or industry. Jacob started thinking in the direction of possible college projects utilizing his learning up to now comprising tension, temperature, and level control. He was impressed with how technology was put to great use in the form of automation to manage level control in industry. Josef was thoroughly impressed by hearing his son's thoughts on wanting to try out these possibilities in college.

Josef, with a smile on his face, told Jacob that their next chat would be on motion control applications, which left Jacob looking forward to the next conversation, which would include *Motion Control – Control, Synchronization, Interpolation of Axes for Accuracy, and Precision*.

5

Motion Control – Control, Synchronization, and Interpolation of Axes for Accuracy and Precision

Planning and executing movement are daily activities that are very important for maintaining schedules and appointments. For example, leaving home using a small vehicle, reaching the metro station, moving from the destination metro station to your place of work, and clocking in at the entry gate needs no small amount of planning and coordination. The first step is to (mentally) plot the time at which you need to be at the different nodes—home, metro station A, metro station B, entrance to work—and the distance between these points and therefore the speed at which you need to move in the different stages of the daily commute. Depending on the speed, you would choose the appropriate mode when setting off from home, such as walking, cycling, or motorcycling. Then you can plan the speed at which you need to move—how fast you must walk or drive the bike.

In the manufacturing industry, a similar scenario exists. Whether it is a conveyor, robot, mixer, cutter, or roller, they all have motion components such as motors and drives connected to these elements. Electronics together with mechanics form a complete motion system. The objective of any motion system is to move a load to the desired position, at the desired speed, along a specified path. Many times, multiple movements take place in parallel, and these movements need to be synchronized. By synchronization, we mean that the load objects must arrive at a defined location simultaneously. This means that the movement of one object must continuously be in accordance with the location of the other object. Over 75% of applications in machines or factories have motion control elements, which must be synchronized and interpolated in order to handle products. Precise positioning and movement are essential for having a consistent and accurate product. If motion control fails, then it might lead to wastage or losses, or at worst, a lot of damage. When making products such as bottles, wire mesh, yarn, cloth, metal parts, capsules, or syringes, motion control is deployed to good effect. The control challenge is keeping the motion along the planned path and at the planned speed in synchronization with the movement of other objects. The measurement variables in these applications are movement ratio and synchronization.

In this chapter, we will look at the products and control challenges in tandem and look at the effect of not handling motion control correctly. The solutions we will observe in this chapter will be to use drives to control motors, such as servos, steppers, and **variable frequency drives** (**VFDs**). The following are the topics in this chapter:

- The need for motion control
- Control challenge (motion control)
- Overview of an application
- Overcoming the control challenge (motion control)

The need for motion control

Jacob was watching a figure-skating competition being broadcast on television. The grace with which the leading pair were executing the moves was so pleasing, even mesmerizing. The precision with which the skaters were executing the moves so that they could reach the same position on the ice even though they were approaching from different angles, and immediately breaking away onto different paths, was breathtaking. Jacob noticed that the two skaters were executing different moves each time, and tracing different paths of different lengths, yet managed to reach the same position in the correct orientation with respect to each other time and time again! It occurred to Jacob, by now so intent on his voyage of discovery in the field of automation, that here was a challenge that the automation he had encountered so far had no tools to address. He was now eager to have another conversation with Josef. Josef was in the same room, reading the pre-publication copy of a book called *The Art of Manufacturing*, for which he was a designated reviewer. Jacob drew Josef's attention to the TV screen, and they watched the graceful dance, which was like ballet on ice, together for some minutes. Josef guessed that this would lead to some questions pertaining to automation from Jacob.

Jacob launched his attempt to quench his thirst for knowledge. Josef quickly switched off the television and asked Jacob to take a seat on the same couch. Jacob started to explain how he was so amazed by the movements, and in particular how the skaters could coordinate with each other so perfectly. Josef smiled and said it did look as if one skater was leading the movements and the other was following. However, the synchronization was heavily reliant on the music, and both skaters knew where they should position themselves at different points in the music. It was as if the music was a third skater, and the two skaters were following the moves of this virtual skater. However, the two humans needed to keep track of each other to make small adjustments so that the dance was pleasing. This was indeed an insight for Jacob, who could now *see* the invisible dancer making the moves. Now, without elaboration, he could visualize how a large team of a dozen dancers could synchronize themselves, because they had one primary axis, which is the music.

The question that arose was whether this applies only to sports or whether any industrial applications need this concept. Even before Jacob finished asking the question, he noticed a broad smile on Josef's face. It is hard to think of any manufactured product that does not undergo some application of motion control.

Josef set out to clarify some topics and explain how motion control works in machines. What would happen if you applied an emergency brake connected to your rear wheel while riding your bicycle at a decent speed? Jacob instantly responded that the cycle would skid as the bicycle does not have any kind of **anti-lock braking system** (**ABS**). Josef responded that it was indeed correct and the imbalance in the movement of the two connected wheels leads to skidding. What would happen if you were winding thread on a big reel? If the speed of the reel winding the thread at one end is higher than the speed at which the thread is being wound from the other end, then there might be a chance of the thread breaking, or if the speed of winding is slower, then the thread being wound could sag. The reason for this phenomenon is owing to the fact that winding and unwinding have separate motors and both these motors need to be synchronized in order to avoid tearing or sagging. This is seen when a friend is winding the thread on a reel when flying a kite. If your friend is too fast or too slow, it results in errors. This can be seen in *Figure 5.1*:

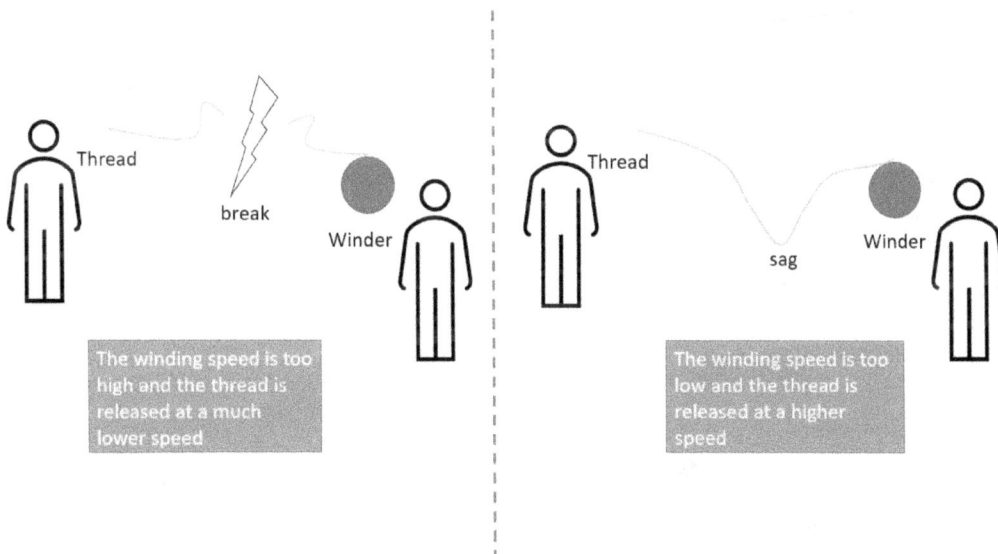

Figure 5.1 – Explanation of how bad synchronization affects thread winding

Errors of bad synchronization are also observed in industry. In industry, these errors usually lead to loss of material, wasted time, machine stops, and loss of productivity.

This chapter helps you to understand the need for motion control. You have had an overview of why motion control is essential in industry. If motion control and synchronization fail, they affect the quality of the product or even damage it. This at times might lead to great damage to machines, or even loss of life.

What actually needs to be controlled?

We think of moving an object along a designated trajectory. The object could be the product that is to be manufactured. It could be the tool that works on the product. It could be that the tool and the product are both in simultaneous motion. As an example, you might think of a product moving on a conveyor belt, and a tool working on the product while it is in motion (perhaps fixing a cap on a bottle).

Which aspects of motion are to be controlled? Fundamentally, it is the position of the tool or product at any point in time. So, the parameters to be controlled are as follows:

- The position
- The velocity (rate of change of position)
- The acceleration (rate of change of velocity)
- Keeping the rate of change of acceleration (known as **jolt** or **jerk**) to a minimum

In *Figure 5.2*, we can see that if you subject an object to sudden acceleration and sudden deceleration, as seen in the second graph, the velocity changes. However, with a steep rise in acceleration, the object experiences a sudden jerk:

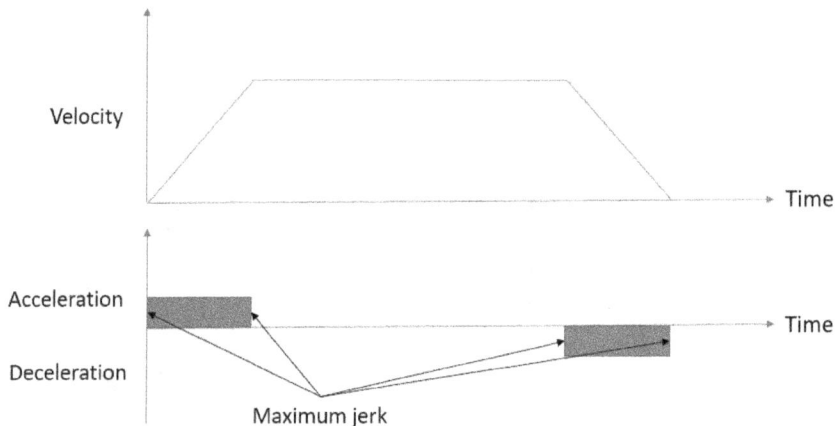

Figure 5.2 – Motion profile with maximum jerk owing to fast acceleration and deceleration

A steep rise in acceleration and deceleration is not a good thing to implement in automation as it might lead to higher wear and tear or even damage the mechanics. You can imagine sitting on a train and the train starts suddenly. I am sure you must have experienced a sudden jerk when the engine suddenly starts moving.

In *Figure 5.3*, we can see that if you subject an object to a gradual acceleration and deceleration, as seen in the second graph, the velocity changes again. However, as it is a gradual rise in acceleration and deceleration, the object's velocity changes gradually and does not experience sudden jerk. Thus, a gradual rise in acceleration and deceleration is typically a good thing to implement in automation, as it reduces wear and tear and prevents mechanics being damaged:

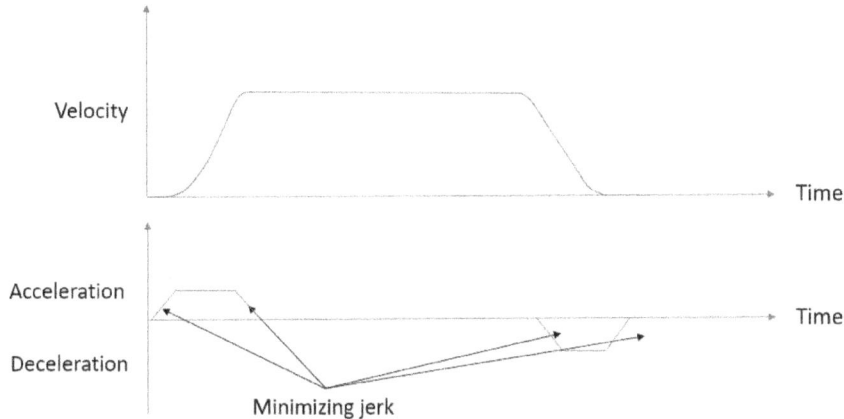

Figure 5.3 – Motion profile minimizing jerk owing to slower acceleration and deceleration

This is the essence of motion control of a single body. As we noted earlier, it is necessary to have several objects moving simultaneously and in a coordinated manner.

So, we visualize each body (product or tool) as one motion axis, and we can make the movement of one axis trace the other axis (which can be called the **primary axis**). The movement of the **secondary axis** can be offset from the primary axis by a time delay or can be magnified by some factor. For example, when the primary axis moves x units, the secondary axis moves by n^*x; the multiplier could be greater than 1, which magnifies the movement, or could be less than 1, which reduces the movement. Imagine robotic arms that trace the movement of the arms of a scientist who makes the motions while looking through a microscope, and the robotic arm performs the same operation on a microorganism mounted on a slide under the microscope.

> **Note**
>
> Primary and secondary axes were until recently called master and slave axes. However, due to their unsavory connotations dating back to the period of colonization, standards organizations have decided against the usage of these terms in motion applications. We have used primary and secondary in this book, but there are various alternatives, such as controller-responder or primary-replica. You can read more about the history behind the original terms and replacements at `https://cdm.link/2020/06/lets-dump-master-slave-terms/`.

In this section, we introduced you to the need for motion control. We introduced topics such as synchronization and the effects if objects in common motion are not synchronized together. We explained topics such as jerk/jolt and what happens if systems are not implemented perfectly. We then explained topics such as the primary and secondary axes, which form the basis of all synchronized motion control applications in industry.

Control challenges (motion control)

Motion control in industry is not merely a single challenge. We focus on two topics now: one is the selection (also called **sizing**) of the motion components, and the second is the control of the motion path itself. Motion components are primarily the motor, the drive (which also may be called an amplifier), and the power train, which consists of coupling and gears. The selection of the motion components (electronic and electro-mechanical) needs to be managed effectively. If the selection of electronics is wrong, then the electronics are unable to sustain the mechanical load and trip out, or if the electronic component is too powerful for the system, it leads to unnecessary expense.

When a motor and a drive unit (or the propulsion unit) are running without anything connected to them, this is called a no-load condition. In this situation, the motor draws its rated current. If the motor is connected to a load via a mechanical coupling or a gear, then the motor draws more current than when it is in a no-load situation. This is analogous to when you are riding a motorbike or driving a car. When it is in neutral, the engine is running idle. However, the moment you engage the gear and release the clutch, the engine gets coupled to the load, which is the vehicle, and the engine's RPM will drop, which you can detect by the change in the pitch of the engine sound. At this point, by giving more gas, you help the motor to sustain the load instead of stalling. This discussion pertains to an internal combustion engine; for an electric motor, different principles apply.

The following diagram shows the setup of a motor with load and without load:

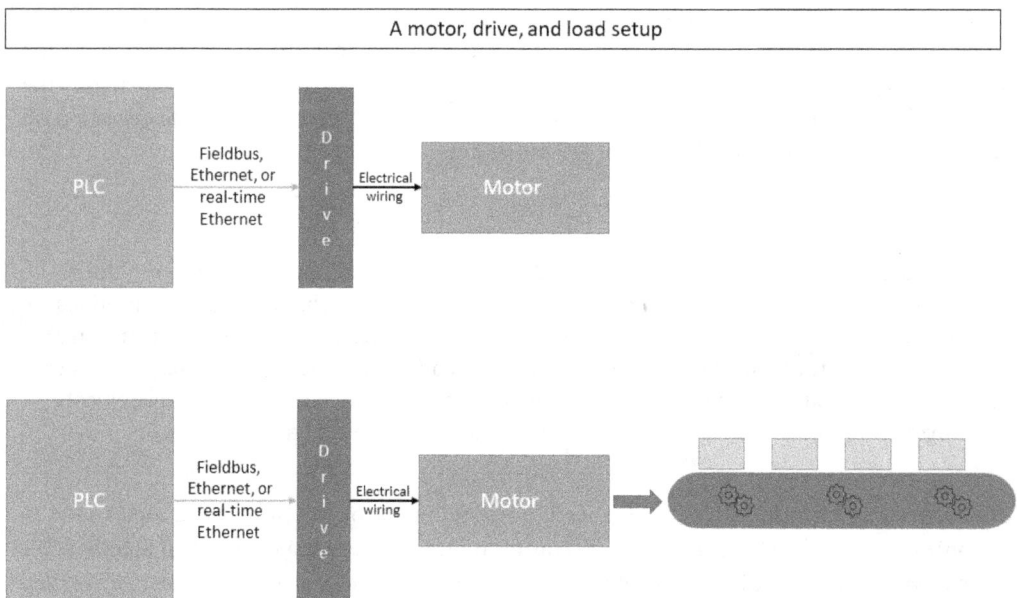

Figure 5.4 – A typical setup of a motor with load and no load

The diagram shows a conveyor connected to a motor. To drive with full load, the motor might draw a certain amount of current. If the load is increased, the motor will need more current. Here, both the motor and the drive are in a stressed situation. If the drive selected cannot supply large currents, then it will trip, stopping the entire machine or the process. If the drive can supply a large current, then after some time, the motor will overheat if it is drawing more than its rated current all the time, and, perhaps in a worst-case scenario, could even ignite.

If the motor is under-rated and the drive is as per the desired rating, then the drive is able to supply a higher current; however, the motor cannot handle the supplied current and thus overheats and trips. If the motor is selected as per desired, i.e., as rated for the application, and the drive is under-rated, then the motor can accept a higher current for providing the right torque to the application, but the drive is unable to supply the desired current and it trips with error. The machine will only run if the motor and the drive are selected as per the desired ratings needed for supplying the necessary torque to the application.

This is provided as an overview in the following table:

Motor selection	Drive selection	Result
Under selection	Correct selection	Motor overheats or trips
Correct selection	Under selection	Drive trips with error
Correct selection	Correct selection	Healthy machine operation
Under selection	Under selection	Motor overheats or trips

Table 5.1 – Correlation between motor and drive selection and its results

The selection of motor and drive needs to be managed at the time of calculating the maximum load on the motor and drive and during the selection of the control system. If there are any errors in the selection, such as the maximum load exceeding the calculated load, it leads to either damage to equipment or lost project time and missed deadlines.

What are the important components of a motion control system? We have the motor as the centerpiece because the motor converts power to motion. The motor is fed with power by a drive (amplifier), which can vary the power supplied to the motor based on the load demand. There are different types of drives with different capabilities. The load is connected to the motor by power transmission equipment such as a belt or pulley, a gearbox, a ball-screw, and so on. Each coupling arrangement is suitable for specific movements and applications. The second control challenge to which we need to pay attention is the movement of the load object. Three aspects are important here: the position of the object, the velocity of the object, and the acceleration of the object. Each of these factors needs control, and each one influences the others. For example, if you would like to reach a target position quickly, you could use high speed. However, on reaching the position, you might have to decelerate sharply and exceed the acceleration limits. To visualize the process, imagine a thick liquid such as yogurt being moved on a conveyor belt. The tub that contains this yogurt needs to stop at the labeling station for the label to be applied. However, if the tub is accelerated sharply at the beginning or decelerated sharply at the labeling station, it can cause a spill. Of course, we would like to move the tub at the highest speed possible to increase throughput. An everyday example could be a lift in your apartment block, office, or college. Obviously, people would like to reach their destination floor in a short time. However, a large acceleration at the start or sharp braking at the destination is unpleasant. There is one more parameter, which is the rate of change of acceleration, called jolt, as we discussed earlier. As you know, people will be uncomfortable if they experience a jolt in a vehicle in motion.

The motion control challenge is to manage the speed, bearing in mind that the number of pieces to be handled (**throughput**) has to be maximized, keeping acceleration and jolt levels within permissible limits. The trajectory is then called the **motion trajectory**. If there are multiple loads moving in tandem, this calls for synchronization. Each load is called an **axis of movement**, and multi-axis coordinated motion control is a nice challenge. You can see that it is not just programming the trajectory; the trajectory influences the selection of motion components and vice versa.

As explained at the beginning of this section, there are multiple options in motion control, and this can be an initial challenge, that is, choosing the appropriate solution for your system.

The following diagram shows a conveyor being run by a **VFD**. A VFD is usually deployed for a conveyor application as it is energy efficient and useful when precision is not of primary importance. Moreover, it is also deployed when the motion is usually in one direction:

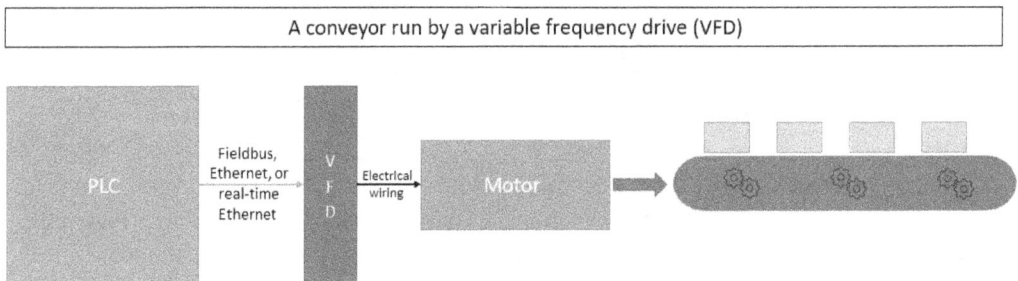

Figure 5.5 – A conveyor run by a VFD

A servo motor and a servo drive are deployed when the application demands high accuracy and precision. A user cannot demand accuracy and precision from a servo using a VFD. Servo drives and motors provide the application with its high acceleration and deceleration needs. They are able to follow the exact motion path, whereas a VFD provides velocity control.

An induction motor reduces speed (**RPM**) as the torque delivered (in other words, **load**) increases. This drop in speed as load increases is called **droop**. A graph that depicts this relationship is called the **speed characteristic of the motor**. A servo motor has a flat speed-torque relation, which means that at any speed it can deliver the same torque. A servo motor is, by construction, a **brushless DC motor (BLDC)**, where the magnetic field is provided by powerful permanent magnets.

Figure 5.6 describes a motion application such as a machine that is responsible for labeling bottles (beverages, medicines, etc.). In this case, instead of using a VFD or an induction motor, a servo is preferred, as the conveyor acts as the primary position, and the labeler follows the primary axis to label the bottles based on the position received from the primary axis. This works in tandem with the bottle presence sensor on the conveyor. These labeling applications are operating at high speeds, which results in the need for higher synchronization among all the motion components. In this case, a synchronization between the VFD and the servo motor. Moreover, it needs a high degree of precision and accuracy for labeling, hence the use of servo motors and drives:

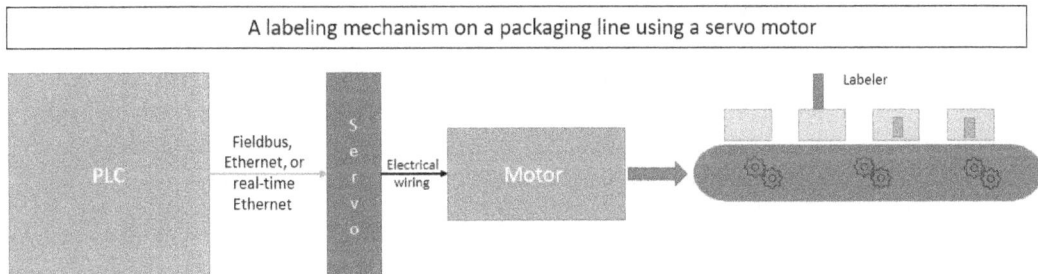

Figure 5.6 – A labeling machine run by a servo drive and motor

A stepper motor is designed to move incrementally and has a higher number of poles. The stepper motor moves a specific number of degrees and does not need feedback mechanisms, that is, it does not need encoders. They are inexpensive; however, they might lose most of their torque at high speeds.

Another application in the packaging industry is explained in *Figure 5.7*. A capper is a machine that is responsible for capping bottles (beverages, lotions, shampoos, etc.). The capper needs torque because high pressure might break the caps, tubes, or bottles. Thus, in such applications, a stepper motor is deployed for better performance. A typical capping application is depicted in the following diagram. This utilizes the power of stepper motors and drives the servo motors. Moreover, this setup also reduces the cost of the system:

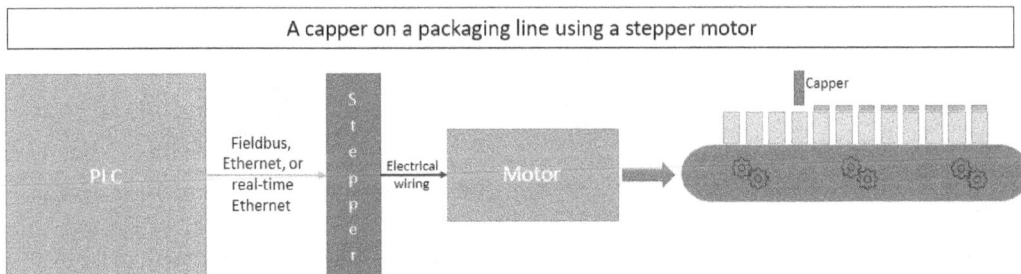

Figure 5.7 – A capper application using a stepper drive and motor

Thus, all motion control components have their own set of pros and cons. The selection of a particular motion component depends on the choice of application.

Another control challenge is to ensure that the motion control algorithm works in the machine together with the mechanics and the assigned interlocks. There might be the possibility of a curtain sensor or two-hand controls to prevent operators from entering unsafe areas. These interlocks should be taken care of by the developer during programming. In *Figure 5.8*, we highlight a two-hand control. The person (operator) needs to load the object (or **job** as it is referred to in the automotive industry) for pressing manually. If the operation of the machine is automatic, then there might be a chance that the machine operates even when the person's hands are loading the object. Thus, to avoid mishaps, the machine is designed with two-hand control. So, the operator needs to load the job, and then, using both hands, press the buttons to start operations. Once the buttons are pressed, the press comes down and presses the job and gives it the required form. The program in the **programmable logic controller (PLC)** is built in such a manner that only when both the buttons are pressed (two inputs are received by the PLC) can the operation start. In addition, the operation will only continue while the buttons are pressed:

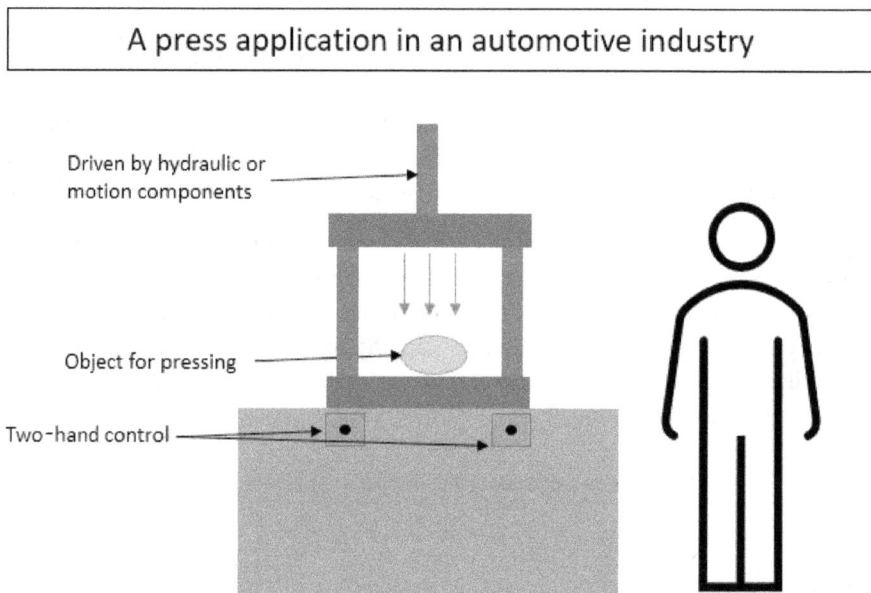

Figure 5.8 – A press application in the automotive industry with two-hand control

Another challenge is when multiple motors and drives are deployed and they need to work in tandem with each other or perform actions one after the other or together. We will see a detailed example of such a synchronized movement in the upcoming section, *Overview of an application*. The application is in the packaging industry, where chips or other such items are packaged. The application for such packaging is a **vertical form fill seal (VFFS)** machine. In this application, the motion components need to pull the packaging material, seal it at the bottom, then fill the packet with the desired quantity of material (chips) and then seal the upper end. All these actions are performed using cams, either mechanical or electronic. Electronic cams need the motion components to be perfectly synchronized with each other. Cams will be detailed later in this chapter in the *Overcoming the control challenge (motion control)* section.

In this section, we introduced you to the concept of motion control in a machine. You should now understand the various possibilities in motion control for various applications and how to select motion control components. You also learned about the other selection challenges and synchronization challenges in industrial applications.

Overview of an application

The most frequent applications where motion control plays a central role are of two types: the first is moving the product or workpiece to the workstation, and the second is the controlled movement of the tooltip that operates on the workpiece. To produce a product, several operations might be needed. Jumping forward to the application example of a VFFS, a plastic film needs to be pulled and shaped, then filled, sealed, cut, and so on. On the other hand, one operation that happens frequently is cutting a metal object to form a specific shape. We are talking about using a lathe, milling, grinding, and so on. The desired shape could be quite intricate, so the cutting edge of the tooltip might need to follow a complex path. It is sometimes even necessary that more than one tool needs to participate, either one after the other or in parallel. Let's look at a real-life concrete example now. Let's explore a VFFS machine application in the packaging industry:

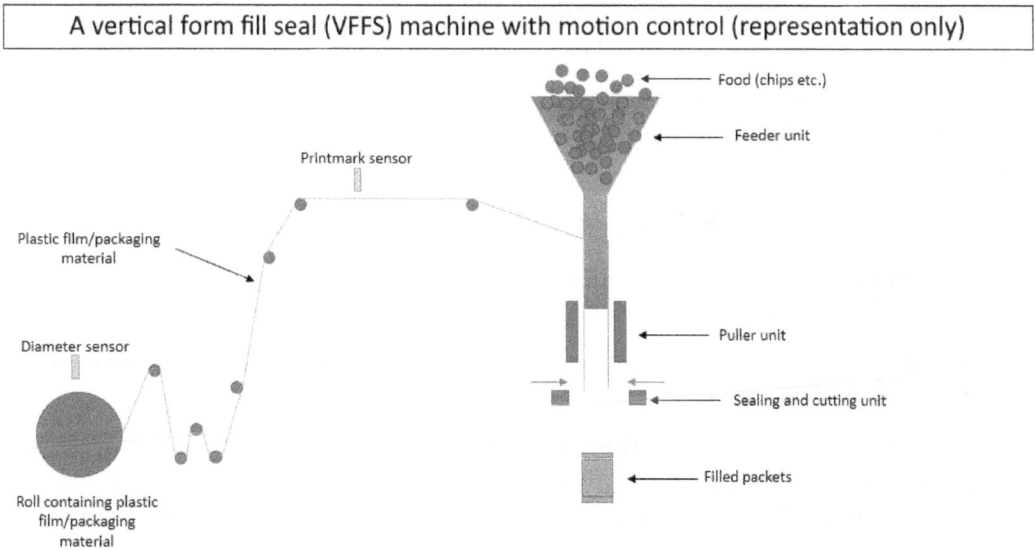

Figure 5.9 – A VFFS machine in a packaging industry

In the preceding diagram, we can see a typical VFFS machine. As the name suggests, the machine stands vertically with the food and packaging material moving from top to bottom. *Form* and *seal* indicate a plastic film or packaging material taking the form of a packet where the film is sealed vertically and horizontally. Meanwhile, *fill* indicates the process of filling the formed and sealed packet with the product, such as rice, wheat, or chips. As packet forming, sealing, and filling are processes in the VFFS machine, these are not shown in the preceding diagram as there are many overlapping areas.

The system is fairly complex, with up to seven motion components (drives) in the machine. The roll containing the plastic film/packaging material needs a servo motor to unwind the roll. The puller unit needs a servo to pull the unwound plastic film/packaging material. The foil feed roller also needs a servo or a VFD-based control. The vertical sealer also has temperature control, and the horizontal sealer can be a hydraulic or servo-/VFD-based sealer.

We saw the role and function of a temperature controller in *Chapter 2, The Art of Temperature Control*, and we saw the role and function of a diameter sensor in *Chapter 3, Tension Control – Managing Material Tension*.

The film or the packaging material is a rolled sheet of plastic with the appropriate branding of the manufacturer. The cylindrical shape is given to the film by the cylindrical machine section below the feeder unit. However, the film sheet has a cylindrical shape and needs to be sealed vertically before filling. The film is vertically sealed by the vertical sealing unit. Now, the base of the packet needs to be sealed horizontally before food is filled into the formed packet. The horizontal sealer seals the base of the packet. Now the packet is sealed vertically and also horizontally at one end. The feeder unit is then allowed to fill the packet with the desired weighed quantity, which might be 10 grams or 10 kilograms. Once the filling is done, the horizontal sealer operates again to seal the packet at the top, while at the same time sealing the bottom of the next packet. The film is continuously pulled and sealed, and packets are formed and filled. All these actions take place simultaneously.

The functions of all the motion components need to be synchronized. Even a minor lag could lead to packets being damaged. The speeds of these machines could be anywhere between 200 to 300 packets per minute.

Any VFFS machine works in the following way:

1. *Accurately weighing products*: VFFS machines are designed to accurately weigh or volumetrically fill pouches with ingredients.

2. *Forming pouches*: The machines are needed to form pouches from thin plastic films. These are either servo or VFD controlled.

3. *Filling desired ingredients*: After weighing products, they need to be filled into formed pouches.

4. *Sealing and finishing*: High product and packaging quality is essential in this section, and vertical and horizontal sealing ensures the quality of the formed pouches.

In such an application, the users and the machine builder expect all the packages to be the same size. In addition, the brand, the user, and the machine builder expect the branding to be visible as defined by the brand. All these aspects of the application need the motion components to work seamlessly together. The size is determined by the length of the packaging material, which is pulled from the infeed roller, and the point at which cutting and sealing happens. But just ensuring all packets are the same length is not sufficient. The packing material has some text and graphics—text to denote the product, manufacturer, and other statutory information and graphics to make the package attractive when it is waiting on the shelf to be picked up. So, it is not just the length of the packaging material but also the pre-printed information that must be aligned. This is achieved by printing a marker called a print mark on the plastic so that with reference to the print mark, and not with reference to where the previous cut happened, we achieve more accurately dimensioned packets that are also pleasing to look at.

The biggest challenge in a motion application like this is to ensure all the motion components work in tandem with each other. We will see how the motion challenge is overcome in the upcoming section.

In this section, we provided an overview of a VFFS machine and the number of motion components involved in the manufacturing of a simple packet of rice, wheat, or chips. You should have also realized that working with motion is a complex affair and a lot of effort is required to perfectly synchronize these electronics and mechanical components.

Overcoming the control challenge (motion control)

As we study the scheme of a VFFS machine, we realize that the machine basically converts a piece of flat film into a *bag* filled with goods. The process of conversion involves two axes. One is the puller axis, which is aligned with the direction of movement of the plastic film; the second is the cutter axis, which is perpendicular to the direction of movement. The cutter axis must move at a fixed ratio to the movement of the puller axis. The puller axis draws the required length of foil or film, then the cutter operates and the sealer seals to provide a closed bottom to the new bag. Then the bag, which is open at the top, is filled with the material. After this action, the puller pulls the packet so that the open end at the top can be sealed by the sealer. Of course, the top end of this packet is the bottom end of the next packet.

If we plot velocity versus the time of the movement of the puller axis, it will look like a trapezoid, with inclined sides and a flat constant velocity pull. The sum of the time for the flanks and the flat area of the trapezoid describing the velocity profile determines how many packets the machine creates per minute. This appears relatively simple, but this is a rather simplified description.

To paraphrase, we need to control the position of the film and the position of the cutter/sealer in a synchronization. To do this, we assign a motion trajectory to both of these axes, and this trajectory is repeated for each bag of goodies. The trajectory must be plotted in such a way that the cutter reaches the foil at a particular point on the film and at a particular time. The position is determined by the length of the pull, and the time is determined by the time taken to fill the bag with the product.

As you know, in real life, there are complications. We want to make as many packets per minute as possible. So, we try to make the sides of the trapezoid of the velocity profile steeper, which means we increase the acceleration and deceleration. So, we get higher speed, but with some serious disadvantages. If the axis, which also means the big heavy mechanical parts, is accelerating and decelerating, it creates noise. Apart from noise, this sort of movement creates vibration and increases wear and tear on the machine parts. So, we do not like steep flanks much. A better design would be to make the flanks like S curves (refer to *Figure 5.2* and *Figure 5.3*).

Beyond that, some time is consumed at each step. Filling a bag with material takes time, which depends on the material. For example, if potato chips take some time, material such as rice grains could fill the bag more rapidly. An item such as common salt could fill its bag quickly, but if the air is humid, it can lead to clumps. The next complication is the film's material and thickness. Different products and manufacturers use films of different thickness, which take a different amount of time to seal. Of course, package sizes vary.

In order to make a faster machine, we can try to make the sealing axis move while the puller is still moving. This saves time compared to the sealing axis waiting for pulling to stop and then starting the journey. More sophisticated designs involve the sealer moving along with the material and sealing while the material is moving.

So, what does this mean for the design of motion profiles? In this case, we deploy a virtual axis, which is set up at a constant movement as desired in one direction. As it is a virtual axis, there is no physical product such as drive or motor associated to this axis. All other motion components are synchronized to this virtual axis. The two secondary axes—puller and sealer—are linked to the primary axis using an electronic coupling. In mechanical systems, this coupling was achieved by a mechanical cam; in electronic motion control, we use a software simulation, called an electronic cam. A **cam** is a disk with a profile mounted on a rotary shaft. A follower doesn't trace the circular motion but instead traces the circumference profile, and hence we can generate different types of motion from the fundamental circular motion. The point to note is that for one rotation of the shaft on which the cam is mounted, the follower completes one cycle of its motion, no matter how complicated the trajectory. This is the underlying idea of synchronized movements.

One fallout of the earlier discussion is that the different axes have different distances to travel but must reach the defined destination at the same time. Hence, the secondary axis must adjust its motion by tracking the primary axis to reach the target location. Here, we are talking about reaching a position and adjusting the velocity to reach the position at a defined time. But adjusting the velocity requires acceleration and deceleration, and there are limits to this because we want to avoid excessive jerks and jolts.

How do we create an effect of coordinated movement? If the secondary axis has to replicate the primary axis with a speed adjustment, a gear mechanism can be used. The speed of the secondary axis is at a fixed ratio to the primary axis, depending on the gear ratio. This is to change the speed profile. This is a form of an electronic cam. The S-curve motion profile injects less vibrational energy into the mechanics connected to the motor and thus is very powerful. The speed profiles need to be adjusted for point-to-point movement to reduce jerk, and start and stop the load. If the position profile needs to change, we could use a cam mechanism. With these two tactics, almost all motion problems can be neatly solved. There are very complex machining operations that need the interpolation of multiple axes as well.

In this section, we introduced you to the possibility of overcoming the control challenge of motion using the primary and secondary axes principle. This principle is frequently used in the automation industry and is usually unavoidable, especially when the axes need to be synchronized.

Summary

In this chapter, we introduced you to the need for motion control. We also highlighted the challenges that the industry faces without motion control, as well as the challenges faced by the developers and the machine builders incorporating motion control. Usually, motion control applications are critical, and the developer, as well as the machine builder, need to be careful to ensure various safety measures and consider all interlocks.

After introducing you to these challenges, we explained various parameters in motion control, namely the position, the velocity (rate of change of position), the acceleration (rate of change of velocity), and keeping the rate of change of acceleration (known as jolt or jerk) to a minimum. These parameters are vital for the smooth functioning of motion components and mechanical parts. We then introduced you to a typical complex motion application in the packaging industry—the VFFS machine. This provides a realistic view of the task at hand for all stakeholders, especially in handling motion applications.

At this moment, just like you, it was too much for Jacob to process such detailed and complex information. However, Jacob was happy to see so many topics intertwined in complex machines—such as temperature control, tension control, and then motion control. VFFS machines had all these components. He was now getting a taste of a holistic automation setup and realized that all smaller applications can be clubbed together and can be utilized to build a completely new and complex application that is capable of manufacturing and producing various products that are used in our daily lives. This time it was Jacob who was mentally exhausted with this information and asked Josef to finish for the day.

While closing, Josef did give Jacob a brief idea of the next topic at hand. The next topic was easier to understand than motion. The next chapter will introduce you, and Jacob, to *Material Dispensing Control*.

6

Material Dispensing Control

An instance when you might encounter weight challenges is when you undertake a journey by air. All airlines have a limit on the weight of baggage you can carry on the flight. In this case, we use weight as a measuring parameter. All weighing scales have a strain gauge or load cell to sense the weight and convert it into an electronic signal for further processing, which can convert it to numbers that humans understand.

In the industry, too, we need to measure the number of materials to deliver to the consumer in the right quantity. The quantity of liquids is usually measured by volume – for example, in milliliters or liters. When materials such as oils (edible oils) and lubricants are filled in containers using volume, accuracy and precision when filling their containers are important to a consumer. The control challenge here is managing the quantity of filled material (in a bottle or can). In such applications, the measurement variable is the **volume**, making it a volumetric filling application. In the case of other materials, such as powders, grains, and other solids, **weight** is the convenient parameter. Nowadays, in the case of delivery to consumers, it is preferable to supply them in a prepackaged form. Such packaging ensures that consumer interests are better protected in terms of hygiene and safety, and also in terms of weights and measures.

In this chapter, we will examine the products and control challenges in dispensing material in accordance with the prescribed quantity in terms of weight or volume. Some packets are filled with enumeration (a count of the number of products, such as fruit, etc.). We will highlight various filling applications in the industry and how they are controlled with a PLC. These are the topics we will cover:

- The need for quantity control
- Control challenge (quantity control – weight or volume)
- Overview of an application
- Overcoming the control challenge (quantity control – weight or volume)

The need for quantity control

It was a weekend, and Josef gave Jacob a list and sent him to get the weekly shopping. Jacob took out his motorcycle and went first to the vegetable shop, where he picked up the fresh vegetables and fruit on his list. On his way to the supermarket to pick up the other items on his list, Jacob thought about how the vegetable vendor had sold him the vegetables. He weighed them and then put them in his bag. Jacob realized that the billing was based on the weight of the item. If the vegetable vendor weighs the item incorrectly, adding more items and charging less, the vendor is at a loss. On the other hand, if the vendor adds fewer items and charges more, it is a loss for the buyer. Hence, in either case, one party is at a disadvantage. Jacob realized that weighing is important in this transaction.

Going further, he realized that so many products must be weighed while being packed. He reached the local supermarket and walked along the aisles, where he confirmed his assumptions. Many packaged items were sold based on weight, such as rice, wheat, flour, and so on. However, he was a little taken aback when he walked to the liquid section. He was expecting that all liquids would be measured by volume in milliliters or liters, but he was surprised to see that some liquids, such as juices, were packaged based on their weight. He further noted that these liquids had higher viscosity. He returned home, put away all the items in their rightful place, and then walked over to Josef.

Jacob started narrating his experience at the vegetable vendor and how his thoughts took him to the supermarket, where he observed something weird. Josef understood that Jacob was on the right path, and it was time for the next lecture. Josef was happy that Jacob was already aware of certain concepts of weighing control and its need in the industry. Josef was glad that Jacob could now relate to various real-world activities in industry and automation, even if there are exceptions, such as the weighing instance he had observed. Josef was also happy that he could take Jacob further along on the application of **Form-Fill-Seal**, which they had jointly explored earlier, as described in *Chapter 5, Motion Control – Control, Synchronization, and Interpolation of Axes for Accuracy and Precision*.

This time, Josef didn't need to start from zero as Jacob had already done his research and his inquisitiveness made him understand various aspects of weighing control. Still, Josef provided a brief background before moving on to the control challenge.

In industries such as food packaging and cement, products need to be weighed, especially those that are sold based on their weight, such as wheat, rice, flour, chips, tea, coffee, and even cement. All consumer products must be weighed before being packed into bags and containers, and the labels on the containers must carry the nominal quantity as they are shipped to consumers. Additionally, there are products such as colas, water, liquid medicines, carbonated drinks, juices, and so on that also need to be measured, but all these liquids have low viscosity and are filled by measuring according to volume in milliliters and liters. Thus, the machines filling such liquids are designed to measure the liquid by measuring the flow of the liquid while filling. This is usually done using a **flowmeter**. Products that have high viscosity are not amenable to conveniently being measured using flowmeters. Such products are weighed and filled to deliver the right quantity to consumers. Some everyday products that have a high viscosity are ketchup, spreads, edible oils, lubricants, and so on. Delivering the desired volume of liquid product into a container is simply called **filling**. It is worth remembering that the density of liquid changes with temperature. While for many products this is not so important, it becomes vital in the preparation of some high-value products, such as medicines and perfumes.

In this section, we saw how weighing control is important in the industry. Moreover, we saw how liquids with high viscosity are measured by weight and not by the filling quantity. We will look into the control challenge in the next section.

Control challenge (quantity control – weight or volume)

The control challenge starts from the point the weighing needs to be done, whether it is *on the fly* or *at rest*. **On the fly** means the bottles, cans, and jars are on a conveyor belt and as the containers move along, the weighing is done and, at the same time, they are filled with the product. On the other hand, **at rest** means that the bottles, cans, and jars are stationary while being filled. In both these cases, there is a need for utilizing a strain gauge or a load cell that provides the signals to the PLC for measuring the weight. In both these cases, the products will always be transported by the conveyor belt. The only difference is whether or not the conveyor belt stops while the filling is in progress. The following figure shows a high-level representation of the filling system and its weighing mechanism:

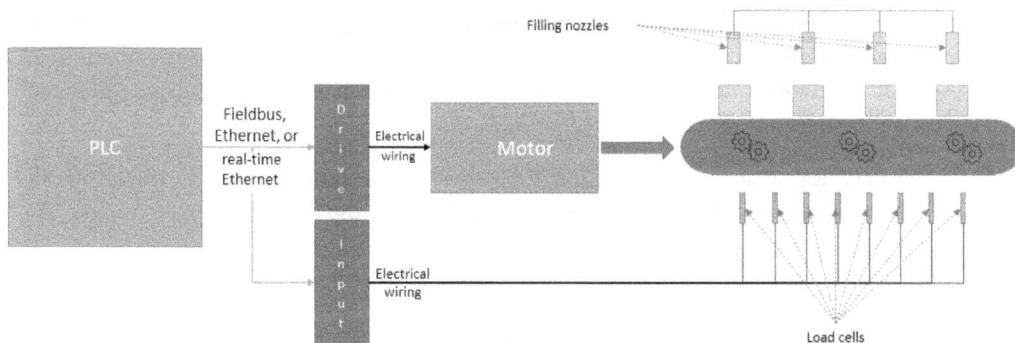

Figure 6.1 – A filling application using a conveyor-based system for transporting products

The number of load cells differs depending on how the weighing takes place. We will elaborate on this in the upcoming *Overview of an application* section.

There are four types of load cells: pneumatic, hydraulic, capacitive, and strain gauge. The pneumatic and hydraulic load cells work on pressurized air and pressurized liquid, respectively, to identify the weight of an object or a product. A capacitive load cell works on the principle of capacitance; with the change in capacitance, the system can measure the change in weight. Typically, an industrial application deploys a strain gauge for measuring weight. Strain gauges are a highly popular type of load cell. In these load cells, the gauge offers electrical resistance proportional to the strain experienced. The relationship between electrical resistance and strain is linear and can be easily converted into a force, and subsequently, into weight. A strain gauge load cell is made of four resistances in a configuration, called a **Wheatstone bridge**. The three legs of the bridge are high-precision standard resistances, and the fourth leg is the variable resistance provided by the strain gauge. Such a Wheatstone arrangement provides accurate measurements. A capacitive load cell is the most sensitive and accurate, followed by the strain gauge load cell. Selecting the type of load cell depends on the sensitivity and accuracy needed for an application.

A load cell is not a simple device to handle. For one, typically, the strain gauge itself is very sensitive, even to minor variations. Hence, the reading has a continuous fluctuation, which can even be due to air currents. A strain gauge is susceptible to temperature variations, so special care needs to be taken to ensure consistent readings. Before taking the reading value, it is advised to allow some time for the stabilization of the counts. But this works against rapid weighing, especially if we aim for a high-speed weighing process, which is necessary for high throughput. A second challenge is the very low signal amplitude from the bridge. This is in terms of millivolts and can easily get distorted by electrical noise. Modern load cells provide an amplification near the source of the signal so that the values can be transmitted over longer distances. But this in turn means a different protection scheme is needed to house the amplifier card away from the PLC cabinet.

Similar to our example in the temperature controller, there are dedicated load cell controllers too. The load cell controller could be a dedicated unit or part of the PLC. An integrated controller would have higher benefits as the load cell control is integrated into a PLC, which provides complete control to the software developer. However, low-cost PLCs might not offer an integrated controller, and machine builders use a dedicated weighing controller and exchange information with the PLC using additional digital inputs and outputs.

Figure 6.2 shows a typical setup in an application; it shows a representation of the setup of a load cell in a real machine that deploys a load cell that is not integrated into the PLC:

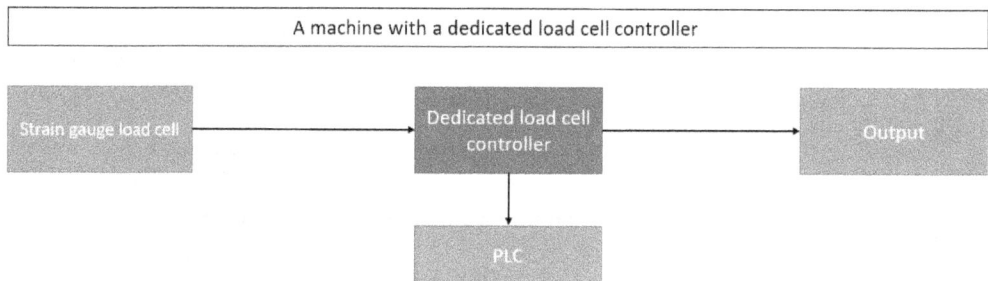

Figure 6.2 – A setup with a dedicated load cell controller and a PLC

In *Figure 6.3*, we show a load cell that is connected to a PLC having integrated abilities to convert signals to electronic data. PLCs have function blocks that process these signals, and the input modules can sense signals received from the strain gauge:

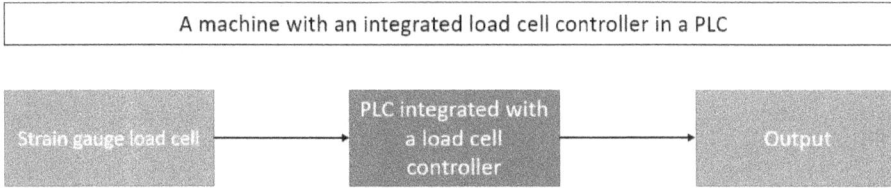

Figure 6.3 – A setup with an integrated load cell in a PLC

In *Figure 6.4*, we show a typical application, which has a conveyor and load cells connected for such filling applications. The filling of the bottles, cans, or jars will only take place based on the presence of cans, bottles, or jars below the nozzle. The load cell provides signals in a 4-20 mA current to the load cell module. Depending on the input received, the load cell converts the values to a corresponding weight. Thus, if a 4 mA signal corresponds to 0 kg and a 20 mA signal corresponds to 50 kg, then for an 8 mA signal, the corresponding value would be 20 kg. This is simple mathematics and can be implemented in a PLC by a developer for recurring calculations. The analog signals are converted to counts between 0 and 65,535; thus, 4 mA would correspond to 0 counts and 20 mA would correspond to 65,535 counts. Such mathematical functions are easily implemented in a PLC for repetitive calculation:

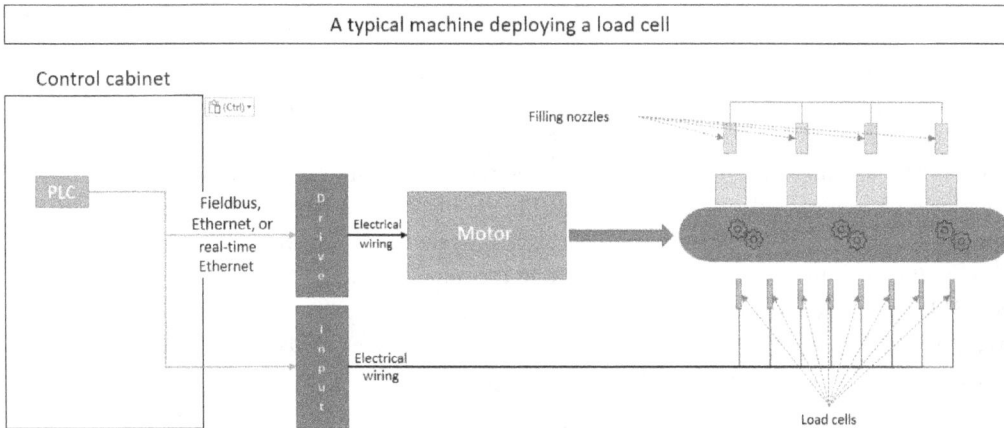

Figure 6.4 – A typical machine deploying a load cell on a conveyor

In this section, we saw an overview of the control challenges involved in measuring the weight of a product.

Overview of an application

Let us take the example of a ketchup-filling application. We can use the same visualization as shown previously in *Figure 6.4*.

Let us look at how such applications are managed. There are several other elements in the machine apart from those shown when a machine builder is building their machine. However, the ones highlighted in *Figure 6.4* and *Figure 6.5* are some of the essential ones. In order to simplify things, let us consider a machine with a single load cell and a single can that is being filled with edible oil or a lubricant:

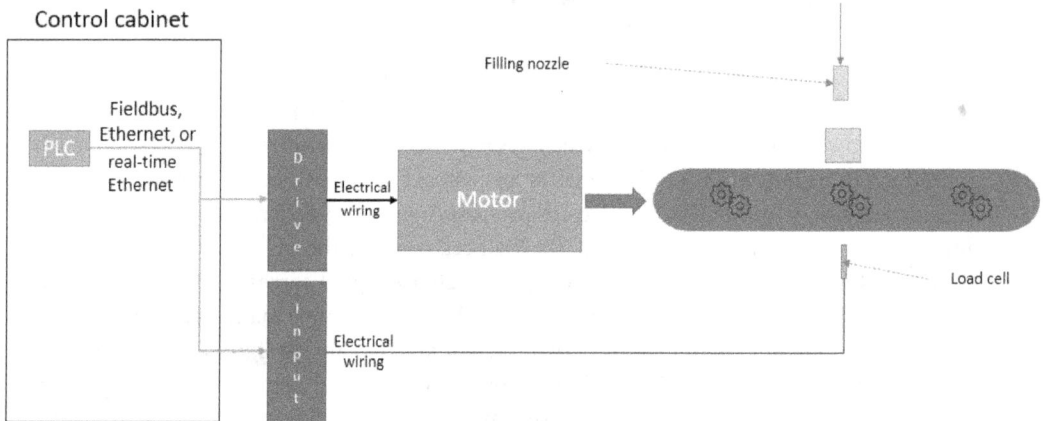

Figure 6.5 – A filling application with a single station

Ideally, an edible oil and a lubricant oil would have a similar consistency, thus making it easier to work with in this example. The conveyor is responsible for bringing the cans to the filling station. This conveyor is fitted with a strain gauge load cell. When the conveyor is moving without cans, the load cells will not be sending any signals, that is, the value at the analog input would be 0 counts or 4 mA, considering it is a module with a 4-20 mA current. There is a sensor to detect the presence of a can below the nozzle, known as a bottle-presence sensor. This is a capacitive sensor, which sends information to the PLC stating that the conveyor needs to stop and the can filling needs to begin only when the can is sensed. These are essential interlocks for the software developer to monitor. The filling nozzle has a valve that opens to start the filling, along with a motor if the liquid needs to be pumped for filling. Once the nozzle opens and the filling starts, the weight of the can starts increasing. It is important to know the specific gravity of the oil to convert kg to liters. Depending on the factory's needs, the machine builder needs to provide the flexibility to be able to switch from kg to liter settings at the click of a button. A software developer would know that this is easily possible to manage in a PLC.

As the filling progresses, the weight goes on increasing. The moment the weight has reached a defined setpoint, the filling needs to stop. It is important to know where the shutoff valve is located. If the shutoff valve is right at the tip of the nozzle, then it is fine to shut off the filling the moment a defined weight is reached. However, if the shutoff valve is not at the tip and there are chances that the oil filling will continue due to gravity, then it is advisable to shut off the filling a bit earlier, thus avoiding overfilling as well as underfilling. The quantity of material that continues to get filled after the feeder valve is shut off is known as **material in flight**. The weight of this material in flight would depend on the density of the material and the quantity in the pipeline between the control valve and the end nozzle. This needs to be calibrated each time during operation by the operator.

In the case that there are multiple filling heads and multiple cans on a conveyor, then the same setup must be multiplied – that is, if there are four filling heads, then the machine would need four can-presence sensors and four load cell modules. Moreover, it is necessary to ensure that the weight of one can does not affect the other can. Alternately, instead of four, if there are only three cans present, then the filling head without a can should not start filling, and if a filling head is faulty, then there should not be any can stopping below that nozzle.

In this section, we saw a typical application with a load cell implemented. We saw how filling takes place on the conveyor and what the interlocks that need to be looked at are in order to avoid wastage and rejections.

Overcoming the control challenge (quantity control – weight or volume)

Unlike the previous motion control and temperature control applications, there are no complex libraries or function blocks that are needed to measure the weight using a strain gauge. It is, however, essential to have a close collaboration between the mechanical and electrical teams in order to overcome this challenge.

While filling high-viscosity fluids, the fluids have higher inertia and there are chances that the fluid continues to flow even after the filling nozzle is shut off. Gravity plays a big role in such applications. Thus, testing such machines is essential to identify the correct filling and calibration of the equipment.

We consider the load cell sending signals between 4 and 20 mA to be mounted to an input module that can sense this input. There will be errors if you have a load cell that can communicate signals between 4 and 20 mA, and if the input module is capable of detecting 0-10V, then the PLC will throw an error:

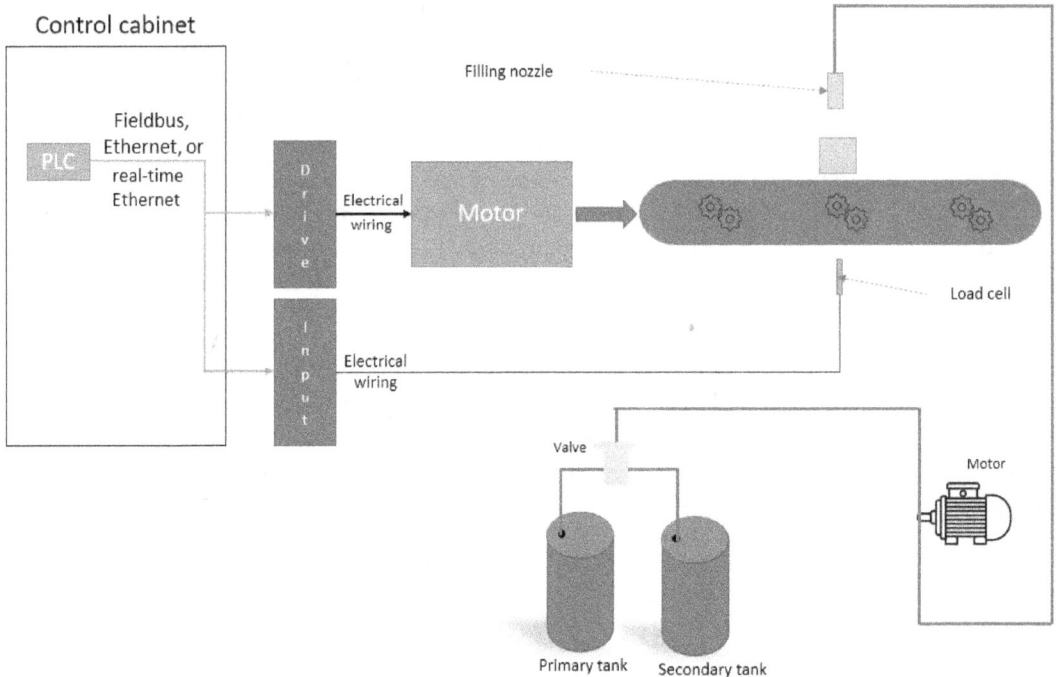

Figure 6.6 – A filling application with tanks (primary and secondary) and a load cell

The close collaboration of electrical, mechanical, and software teams is needed in all applications, but it becomes especially crucial in this case, as it might lead to a lot of re-programming and larger testing times if there are communication gaps. A simple location of the shutoff valve for shutting off the filling might affect the precision of the filling. Just imagine opening a tap to fill a water bucket. Let us assume that the tap is at a certain height and a pipe leading to the bucket is long, say 10 meters. Whenever you turn off the tap, there will always be water in the pipe leading to the bucket. Thus, there will be surplus water flowing in due to gravity or the flow of water. A similar phenomenon is experienced in filling applications. If the shutoff valve is at the tip of the pipe, then the filling accuracy is higher, and if it is further behind the tip, then there might be possibilities for the machine to deliver lower performance and precision. Thus, the machine offering a weighing accuracy of 0.1 kg is different than that offering a 0.001 kg filling accuracy.

In *Chapter 4, Level Control – Controlling Levels of Liquid to Avoid Drying Up or Spilling Over*, we did see part of the application: two tanks that can be switched the moment one dries up so that the machine does not stop operation. This was an engine-filling application. A similar setup can be possible for edible oils, or this setup can be easily replaced with an overhead tank. However, a valve, a motor, and a nozzle would be common in any setup. The developer needs to take care of multiple interlocks as this is an automatic application (i.e., it needs no human intervention), unlike the automotive application where there is a person to insert the nozzle and push a button to start filling. In this case, a bottle or a can arrives on the conveyor, and the process of filling starts. Thus, the developer needs to check that the bottle is present below the nozzle, that the tank is not empty, and that the valve is open before turning on the motor to fill, and then needs to shut it down before they restart the conveyor to take the can or the bottle to the next capping station.

The flow for the packaging industry for any kind of filling is almost the same. *Figure 6.7* highlights a typical flow in the packaging industry:

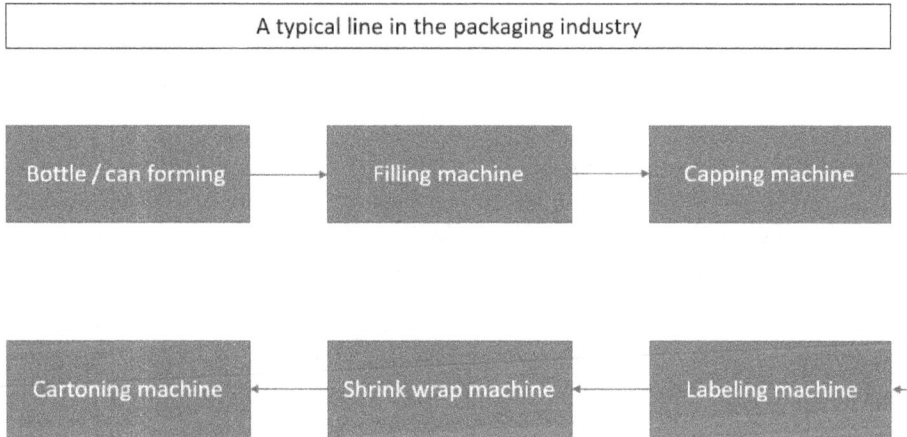

Figure 6.7 – A representation of a packaging line for bottle and can filling

It usually starts with the bottle or can formation moving to the filling area, then capping. Once capping is done, the filled bottles/cans move along for labeling, shrink wrapping, and then finally, packing for dispatch. Instead of filling based on weight, the flowmeter might be used to check the quantity being filled. We have worked on an automotive line where the workers needed to fill two different kinds of oils on two different oil variants. The **Society of Automotive Engineers (SAE)** International grades oil on viscosities. The most commonly used available engine oil would be 5W-30 and 5W-40. Such oil variants are prescribed by the manufacturer while manufacturing a vehicle. Thus, during manufacturing there is a need to select the correct oil variant for an engine. In addition, the person needed to fill one variant using weight and another variant using flow. Thus, one variant had a flowmeter connected to the nozzle, and the other needed to monitor the weight of the product before and after the filling. The parameterization was also different for the two variants. One variant (which was filled using a flowmeter) had all the settings in liters, whereas the other had all the settings in kg. The application is complex, with huge lines of code with a lot of interlocks. However, the application is not mission-critical. The worst that can happen is a waste of oil if the filling goes wrong.

In this section, we saw how a machine builder and a developer overcome the weighing control challenge. We saw two applications, one from the packaging industry and one from the automotive industry.

Summary

In this chapter, we saw some industry examples that need weighing control and the challenges faced by machine builders, end users, and software developers while implementing these weighing applications. We highlighted the need for close collaboration among various teams on the machine-building site and the results of ineffective collaboration. In this application, the reaction times are not too high, yet the machine result is critical. Just imagine the machine dispensing 100 g of extra oil while filling 5 kg. In that case, for every 5 kg, 5,100 g (5.1 kg) is being dispensed instead of 5,000 g. Thus, after filling 10 cans of 5 kg, the end user would have lost 1 kg of oil. Therefore, if the factory manufactures 1 ton of products every day, you can imagine the amount of loss the end user would suffer.

We introduced you to simple weighing applications, such as buying vegetables from a vendor or weighing your luggage while traveling. We introduced you to the concept of filling by weighing and the reasons such an arrangement is essential. We then highlighted the issues that could be faced if such an application fails, and elaborated on an application in the packaging and automotive industries. In this fashion, we were able to give you a brief introduction to the topic of weighing control.

Jacob was able to once again connect the dots. He recollected the tank-filling application and then the motion application and how these two work together with the strain gauge load cell to build a weighing application. He had started with a do-it-yourself kit and was already struggling with the initial concepts; thus, he was fully aware that adding elements will almost always add complexities and will not reduce them. He realized that if any one element in this application was unmanaged, then the entire application would crash. If the filling valve fails to open and the motor starts, the motor trips if the motor selected during machine design is not as per the desired rating (under-rated). However, if the motor is selected during machine design and is not as per desired rating (over-rated), then it might lead to the pipes bursting. If the conveyor fails to stop or the can-presence sensors fail to operate, then the filling starts without a can below the nozzle, leading to waste, or the filling process fails to begin.

After completing six major lessons on automation, Josef was pleased with the progress Jacob had made. It was now time to look into other aspects of the factory, machine building, and automation. These applications provide great insight into automation and the machine-building process. At this point in time, Jacob was looking forward to going back to the do-it-yourself kit and playing with it to develop something totally new and, at the same time, learn about and explore various aspects of automation. He was now looking forward to the next discussion with Josef on *Automation and Humans*.

Part 2:
Automation and Humans

Humans need to interact with automation, giving commands and setpoints, among other things. Similarly, automated systems need to tell humans about the current state of processing. Faults and exceptions need to be made clear. Diagnostic information needs to be conveyed. But beyond these routine communications, there is another level of interaction.

Automation is a faithful servant. Automation is a tool and is built, developed, and maintained by humans. But it is different from all other tools in that it is intelligent. And beyond simply being intelligent, it has now developed the capability of being aware—aware of its surroundings, and even self-aware! In this scenario, humans and automation need to march shoulder to shoulder.

This part has the following chapters:

7

The Interplay of Humans-Machines-Automation

It all began with the wheel. With a wheeled cart, it was much easier to transport a heavy load, rather than carrying the load on your shoulders or the back of a donkey. From such humble ideas, the concept of machines grew. The invention of the wheel was a significant point in human evolution. It was fundamentally different from objects like a lever. The idea was always there, to increase the mechanical advantage; that is, the ratio of the load to be moved versus the effort to be exerted. The determining factor was that the human or the animal exerting the force—the prime mover, in other words—had a limited strength or capability. So, methods were sought which could multiply the force which was exerted by the animal or human. The early means of a lever, inclined plane, wedge, and pulley were intended to change the application of force, either in direction, such as in the case of a pulley, or in magnitude, such as in the case of a lever. Archimedes was impressed by the power of the lever, and he is reputed to have said: *Give me a lever long enough and a fulcrum on which to place it, and I shall move the world.* Human ingenuity did not rest at that point but went further to build more complicated machines. With a wheel coupled to an axle, humans could move more rapidly, using, for example, horse-drawn carriages. It also helped to make tools with pointed edges by grinding the edge of a lance to a sharp point, creating weapons in other words. All these developments in early technology enabled humans to counter the natural advantages of animals; that is, their faster speed and the sharp teeth and claws they possessed. With such inventions, humans rose to the top of the food pyramid. In summary, through the use of machines, humans obtained a unique survival advantage, which helped to obtain dominance over all animals.

In this chapter, we will explain how humans and machines work together. Moreover, we will touch upon how automation is the bridge that brings the two worlds together.

We will cover the following topics:

- Humans and machines

- The need for interaction with machines

- Programming platforms and programming languages

- Data from the process – values, alarms, and events

- Supervisory control

- Mobile first – control from anywhere

Humans and machines

Josef had explained different application examples of automation over the past few weeks. Being a quick learner, Jacob had begun to understand how automation helps to create the desired features in products. He understood that there are many machines to perform various tasks. He also grasped that many machines are controlled by automation. His curiosity, however, did not abate. Many questions arose in his mind: was it always like this? How were things managed before there was automation? Was there always automation? He chose a leisurely Sunday afternoon to engage in a sort-of philosophical discussion with his father, Josef. After all, Josef was his go-to person for such topics that he could not raise at his place of study.

Josef was actually looking forward to a quiet Sunday afternoon to digest his sumptuous lunch. But then, he too had the desire to share his knowledge with his son and orient his son's thinking in the right direction. So, he decided to expand the scope of their inquiry. He told Jacob that before automation there were still machines. Indeed, even today, many machines are not automated. Jacob asked, what is the opposite of automated?

Josef began to explain. The implementation of automation is not binary, that is, automated or not automated. A machine or process can have varying degrees of automation: from completely unautomated to partially automated, semi-automatic to fully automatic. After all, a machine or process consists of many parts, activities, and functions. Any one or all of these functions may be automated, which determines the degree of automation. We can relate this to our washing machines at home. Today, we might visit a showroom and see only top-load or front-load machines. However, a decade ago, there were semi-automatic and automatic machines. In semi-automatic washing machines, there were two buckets, one for washing and one for drying. Thus, human intervention was essential while operating them. After the machine finished the wash cycle, you had to open the lid, pull out the washed clothes and put them in the second, adjacent dryer bucket, and run the drying cycle. Those who were born in the 1990s have probably seen such washing machines. *Figure 7.1* shows a typical semi-automatic washing machine:

A semi-automatic washing machine

Timer switches

Dryer bucket

Washing bucket

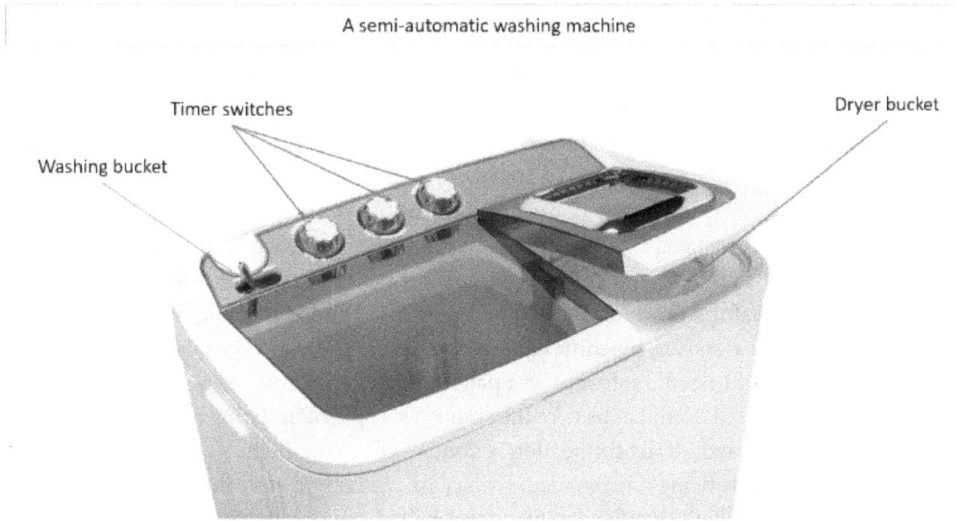

Figure 7.1 – A typical semi-automatic washing machine

A **machine** in our context is an apparatus (or mechanism) that converts a workpiece from one shape to another, which is either ready for dispatch or ready to be further processed by the next machine. A machine with no automation is called a manual machine. Even an automatic machine can be operated in manual mode.

Automation is a faithful servant. Automation is a tool that is built and developed, tended to, and maintained by humans. But it is different from all other tools, in that it is intelligent. But beyond being intelligent, it can develop capabilities of being aware—aware of surroundings, and even self-aware! In this scenario, humans and automation need to march shoulder to shoulder.

Think of the washing machine that we use to wash our clothes, said Josef. Is the machine that we use at home automatic? Jacob answers, hesitantly, that it is automatic, but they need to adjust many settings before switching it on. He mentions how, in the previous week, he got the settings wrong, which resulted in the clothes being damaged. Josef managed to grin, even though it was his favorite shirt that was a victim of this mishap. This is an example of where the amount of interaction and skill needed to operate the machine is high, exclaimed Josef. Even though each individual function has an element of automatic control, the overall function needs inputs from humans. And it is not as if once having set up the parameters, the operator can step away. The machine needs supervision—to check that the water supply is on, the power is on, and there are no weird noises emanating, such as high vibrations or knocking.

The need for interaction with machines

Humans need to interact with machines and hence, with the automation that is behind the machines, to give commands, and to provide setpoints. Similarly, automation needs to tell humans the current state of the processing; that is, the progress made of the task under process. Faults or exceptions need to be alerted. Diagnostic information needs to be conveyed. These are the interactions between humans and machines during the operational state of the machine. But beyond this routine communication, there is another level of interaction. Can you remember the days in the early 2000s when you were trying to download a file via the dial-up connection on your computer? If the connection was bad, there was a high chance that the download would be interrupted and you would have to restart the download. A similar case persists in machines and automation. There are various scenarios where the machine controller might need a software or a patch update. The most likely reason is a feature update or security patch installation. Let us take the simple example of updating a particular piece of software or downloading a patch. If the connection is sketchy, then you might want to keep an eye on the progress while the patch is being downloaded. Otherwise, you might start the download and take a break, just to return and find out that the download was halted, paused, or closed, and you have lost time in the process, needing to reinitiate the entire download.

Interaction while in operation

Let us consider baking a cake in an oven as an example. An oven can have varying degrees of automation. A very basic model, for example, would be operated with on/off switches, rotary switches, and a timer. Some other ovens might have electronic timers and advanced settings. The machine gets powered on with a main switch. A person who is an expert knows how much time would be needed for the cake to bake perfectly and such a person would leverage all the features in the oven. On the other hand, a person baking the cake for the first time might not know how much time is needed for the cake to bake or what the best temperature is to bake the cake. The person would need to keep checking the status of the cake while baking. Thus, the first time baker would need to keep opening the oven and checking if the cake is baked to perfection. If they overbake it, the cake might burn; if they underbake it, the cake won't have the usual consistency. When the first-time baker adds the batter into a baking pan, the person needs to set the temperature and the timer for baking the cake. Now, an experienced baker would set the timer and the temperature to the precise values they know, and then carry on with other activities. However, a first-time basker would set their parameters but would need to ocassionally open the oven and check if the cake is baked.

Here, the operation panel consists of some switches, rotary timer switches, and maybe a lamp or two to indicate operation. Additionally, an alarm lamp to indicate tripping, and so on, is available.

When we deal with an automatic washing machine, the functions and cycle of operation are not different. However, the various functions are switched in and out of by a pre-programmed cam disk. In the modern version, the cam disk is an electronic card. This removes the requirement that a person has to be standing by the machine all through the cycle. The sequence of the operation, as well as the time duration for each step, are available on the cam, and the operator need not select the parameters each time. Coding is available to choose the cycle, depending on whether the clothes are white, cotton, delicate, silk or wool, and so on. The corresponding program then gets into operation. The machine can actually weigh the clothes that have been loaded and adjust the amount of water so that there is no waste.

For most industrial machines, parameter settings and machine variants, which are sometimes called models, are available, such as a printing press or an injection-molding machine, or a **form-fill-seal (FFS)** machine. This is identical to buying a car. The basic model is the cheapest with the bare minimum features. The top-end model of the same car model has all the luxurious features and is priced higher. This is similar to industrial machines, only here there are machine variants and models that offer different functionality. Today, many machine builders brand their machines as **industrial internet of things (IIoT)** ready or Industry 4.0 ready. These machines have the same functionality as normal machines during operation but additionally possess the latest technologies implemented to make them ready for IIoT.

Interaction while the machine is constructed

This interaction occurs during the manufacturing phase of the machine. The machine is manufactured in a factory that is different from the factory where it will be operated. At the machine manufacturing factory, the machine needs to be prepared for the task for which it is intended. On a very high level, you can relate to building a house and staying in one. The only difference is the apartment is stationary; thus, people inhabit it in the place it is built. Now, while the building is being built, there are construction workers, civil engineers, and architects at work monitoring the entire process. Once the work is completed, they move out and make the apartments ready for people to stay in them.

Machines come with several features and options. Depending on the buyer's needs, the features and options are selected. As the machine is being assembled, the mechanical units corresponding to the customer order are mounted. Then, when the electrical wiring is completed, automation hardware is mounted. This hardware, too, will depend on customer order. In the final step, automation software has to be downloaded into the **programmable logic controllers (PLC)**. This software is specific to the customer hardware, according to the model and features that are chosen. This is a level of human interaction with the machine through an automation unit that is mounted in it.

This is identical to the apartment analogy. Usually, all flats in an apartment are identical and include the same areas. In a single building, there could be multiple floors with two and three **bedroom, hall, and kitchen (BHK)** flats. Usually, all two BHKs have the same design and area and all three BHKs have the same area and design. However, it is up to the owner to design the interiors according to their preferences and tastes. Thus, once you enter each flat, they might look a bit different.

Another similar example is buying a car. The car manufacturer has a limited number of variants offering different features. All variants are priced differently. Thus, all cars in a single variant will have the same features. However, after purchasing, the buyer still has options to add some aesthetic features in the form of accessories. Thus, based on the buyer's preferences, they can make certain changes.

Interaction at the design and testing stages

In these stages, the operational program is developed. For this purpose, we need a programming platform. Using this platform, the programmer develops and tests the application program and then tests it with the machine to check its proper functioning.

A tester might write test cases and automate this process, but such an approach is for a software-only solution. However, there could be human and machine interaction to see whether the test results are in line with the requirements or whether smaller deviations are acceptable. Thus, while testing a machine, the software, hardware, and mechanics all are tested working together with humans trying to create faults to check the robustness of the machine. Moreover, the tests are also to identify the quality of the final product. Many companies have a **factory acceptance test** (**FAT**) within testing. Thus, a factory, while buying a machine, will conduct a FAT that the machine builder needs to pass. Subject to this clearance, they can ship the machine to the factory.

Interaction at the supply chain and procurement stages

Whether you are building a machine or using a machine to manufacture a product (like the ones we learned about in previous chapters) there is a great deal of human involvement in the process. Supply chains can be automated but the basic process and management need to be managed by humans. There are concepts, such as **just-in-time** (**JIT**), which help manage and reduce the inventory overhead. However, the entire process involves a huge amount of human involvement; that is, not everything is fully automated and human intervention free.

Factories need to source raw materials for machines to function, operate, and produce goods. These raw materials also define the quality of the product being manufactured.

In this section, we introduced you to the aspects of human involvement with machines and how these interactions are essential for the machines or lines to function as desired. We highlighted four major areas of interaction, namely, interactions while the machine operates, interactions while the machine is being commissioned, interactions during design and testing, and finally, interactions in the supply chain area.

Programming platforms and programming languages

A **programming platform** is a suite of software that provides facilities for entering and editing a PLC program along with a compiler and a debugger. Furthermore, there will be a downloader and sometimes an uploader. Programming platforms are mostly PLC manufacturer specific.

Application software is written in a human-readable form and many choices are available. Previously, there was a plethora of programming languages—every vendor had its own variant of programming language. Therefore, a standard was developed by the **International Electrotechnical Commission** (**IEC**). This standard is defined under clause *IEC 61131-3*.

IEC 61131-3 languages

The majority of PLC systems adhere to the IEC 61131-3 standard that defines two textual programming languages: **structured text (ST)** and **instruction list (IL)**; as well as three graphical languages: ladder diagram, **function block diagram (FBD)**, and **sequential function chart (SFC)**. Some systems also support **high-level languages (HLLs)**, such as C.

Structured text

ST is an HLL and, usually, it is a modified version of a computer programming language, such as Pascal or C. This makes for compact programs, providing multiple instructions in a single line, and is much more programmer-friendly. *Figure 7.2* shows a typical representation of logic written in ST:

```
CASE StepSequence OF
Case 1: If gVar1 <=5 then
          gVar2:= gVar2+1;
Case 2: if gVar1>=10 then
          gVar2:= gVar + 1;
Else
          gVar:=0;
END_CASE
```

Figure 7.2 – A typical representation of logic written in ST

The controller will execute the logic from top to bottom.

Instruction list (IL)

IL is a lower-level language, like an assembly language. It consists of a sequence of instructions to the PLC. It is also called a **statement list (STL)** programming language. *Figure 7.3* shows a typical representation of an IL:

```
LD              gVar1
ST              Timer_Ton.IN
CAL             Timer_Ton(
                PT:=t_sec
                ET=>t_out)
LD              Timer_Ton.Q
JMPC            var2
```

Figure 7.3 – A typical representation of an IL

It consists of instructions in code that the controller executes from top to bottom. A developer needs to write instructions and connect variables to them so that either the input is read from these variables or the output is written into them.

Ladder diagram

Typically, PLCs are programmed using ladder logic, which is used in industrial control applications. These programs resemble a ladder and hence the name. It consists of two vertical rails and a series of horizontal rungs between them. This has a close affinity to the earlier controls, which were based on relay logic and remained a favorite of electrical engineers.

Figure 7.4 shows some typical ladder logic:

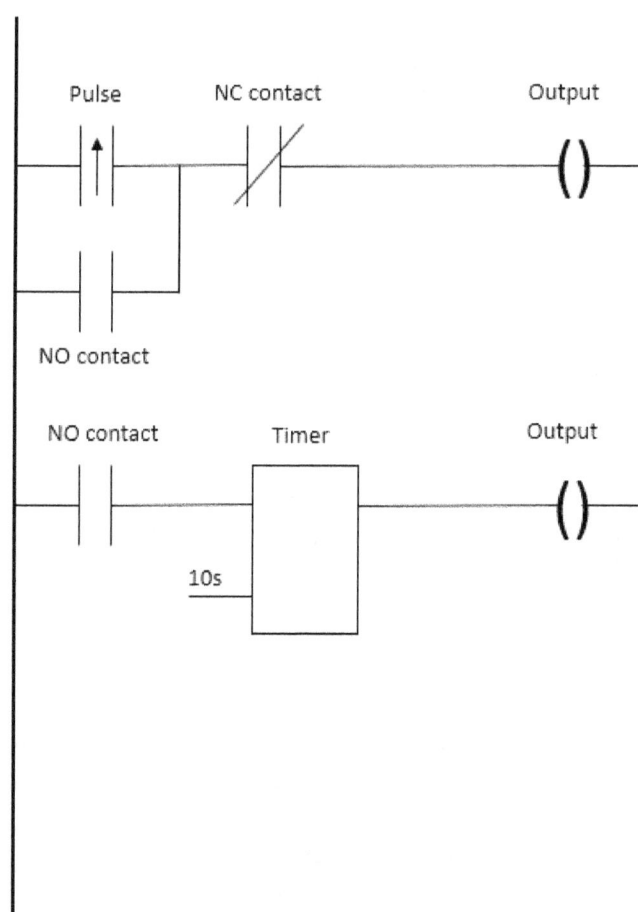

Figure 7.4 – Typical ladder logic in a PLC

There are various elements that can be used to pictorially program a PLC. The preceding diagram only shows a pulse, a **normally open (NO)** contact and a **normally closed (NC)** contact, an output, and a timer. As the name suggests, a NO contact is a contact that does not allow things to pass, for example, a switch for powering an electrical bulb is normally open, hence the bulb is normally in the *off* condition. An NC contact is a contact that, at rest, allows things to pass, which is exactly the opposite of a NO contact.

Thus, in industrial automation terms, a NO contact will not pass electrical signals when at rest. A sensor will switch to 1 when it senses something. When it switches to 1, the electric signal is received by the digital input modules and then any action can be performed. Thus, the sensor input is a NO contact in the ladder. Only when the sensor senses anything will the circuit be allowed to complete.

Function block diagram

FBD provides an abstraction of the earlier languages by enclosing an entire function inside a rectangular block with **input/output (I/O)** connectors. So, a programmer can be more productive, in that big chunks of code can be introduced by a simple block. *Figure 7.5* shows what an FBD looks like in a programming tool:

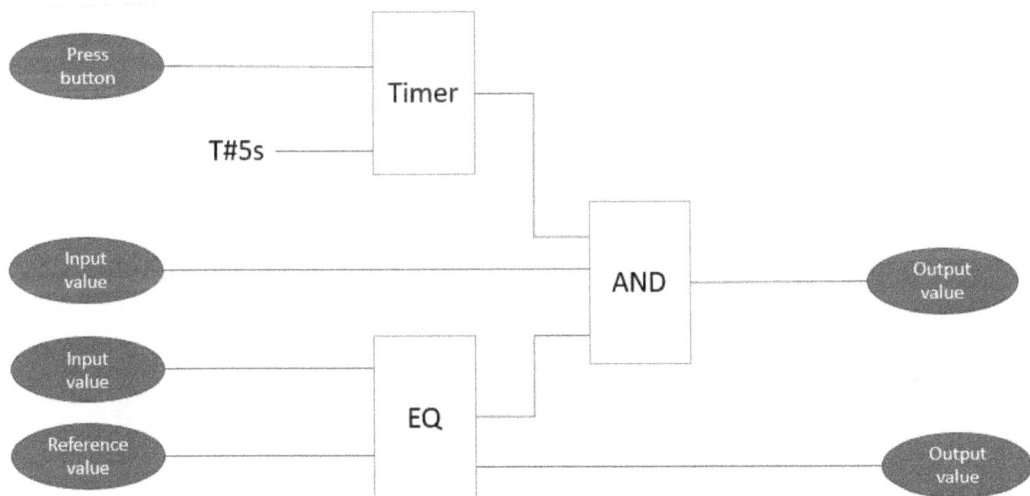

Figure 7.5 – A pictorial representation of an FBD

Every brand has some variation in the look and feel but the basis is that such a programming method is a graphical representation of programming. The logic in *Figure 7.5* can be tweaked and made into a coffee or drinks-vending machine with additional logic. The flow goes from left to right and the developer also needs to work from left to right. The PLC executes the code from top to bottom and left to right.

Sequential function chart

An SFC represents the flow of control in a familiar flowchart form, which makes the logic very visible. However, at this level of abstraction, debugging becomes a little more difficult. *Figure 7.6* shows a typical SFC:

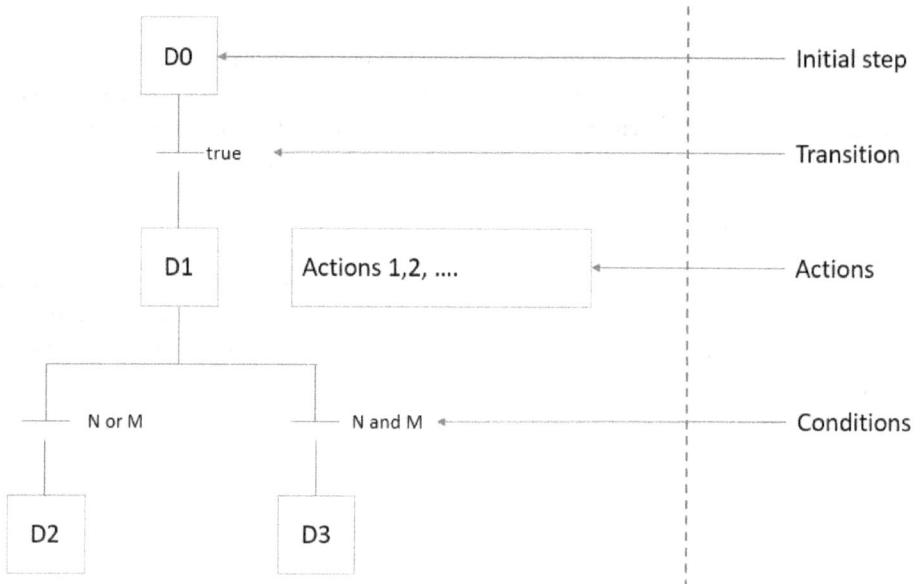

Figure 7.6 – A typical representation of an SFC

As you can see the developer needs to identify various conditions and then sequentially write them. The controller will sequentially execute the code from top to bottom.

Various formats of the application software

Application software, as it is written by a programmer, will be in one of the previous forms, and is available, usually, as a windows file in text form. But, every so often, real-life control solutions are complex and consist of many programs or tasks. All of these tasks are bundled together as a project. This stack is called **source code**.

The software in the form of source code is then compiled to convert it into **object code**. This object code gets *built* into the project for the PLC in a form readable by the CPU of the PLC. This is the download version, which we can also call **machine code**.

Up to this point, we have seen the different programming languages covered under *IEC 61131-3*. However, in the next section, we will now see how the data interacts to build a complete system.

Data from the process – values, alarms, and events

How can humans understand what is happening at any given point in time in a machine? The **human machine interface** (**HMI**) provides all such information in the form of alarms, warnings, and status information.

Elementary operator interface – switches, dials, and meters

An operator needs to know what the machine is doing at any given point. A machine needs commands from the operator. So, every machine needs an HMI.

On a washing machine, for example, there is a small lamp next to the step of the cycle that is running. There may be a rotary switch to command which cycle step to execute. There may also be a power light to indicate that the machine is switched on. There will also be a power switch. In a more sophisticated machine, there may be a small two-digit display to indicate how many minutes are remaining in the current step. There can also be a small LED display to indicate which operation is in progress, and how many minutes till the end of the program. *Figure 7.7* shows a typical washing machine dashboard:

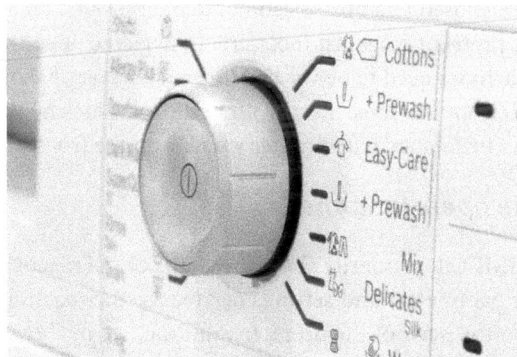

Figure 7.7 – A typical dashboard of a new generation washing machine

Humans can interact with the machine to adjust it to the appropriate settings.

HMI and operator panels

If we talk about a more complicated industrial machine, such as the **injection molding machine** (**IMM**), there is a bigger display panel connected to the PLC. The amount of information displayed is much greater.

There is a limit to how much information can be accommodated in one screen from the perspective of operator comprehension and readability. Hence, the HMI would have information spread over many pages or screens. Each page would have information that belongs together. There may be some information which is displayed on every page, such as date and time, and some important parameters, such as production count.

Diagonal sizes, colors, and touch

Displays come in many sizes. This can be said of mobile phones, where you can choose between various sizes all the way up to tablets. HMIs in the industry also have various sizes. They could have metal or plastic bodies.

Typically, HMI sizes are indicated according to the diagonal length. Typically, the HMI screen sizes on machines in the industry are usually from the following possible sizes: 5.7", 10.4", 15", 19", and so on. HMI touchscreen displays are nowadays always color. There are touch panels, where switches are replaced by sensitive elements in the display. Keys are needed for manipulating the display itself, such as selecting the screen, paging, and so on. They are also needed for operating the machine, to trigger a function, for example, to clamp forward.

Ingress Protection

The HMI is an operator's window into the machine. The HMI is exposed to the atmosphere on the shop floor where there may be dust and fumes in the air. **Ingress protection** (**IP**) indicates, in general, how well-protected an object is by its enclosure. Since electronics are rather sensitive to heat, dust, and moisture, they are enclosed in a box or cabinet to offer protection against such pollutants. The most important thing is protection against moisture. In hygienic operations, such as in food or medicine production, the machines need to be washed down after every batch. Hence, a high degree of protection is needed in such applications. So, a typical rating for such applications is IP65, which means the operator panel can be cleaned with a water spray from the front.

Topics displayed on the operator panel

The operator panel or the HMI will primarily show the status of the machine. It also provides the option for a person to enter parameters and settings needed for the machine to operate smoothly or to make changes based on the need of the users. In addition, it can show other aspects, such as alarms, warnings, and the health of the machine. If the machine is handling multiple products, then it can also store different parameters, which can be loaded at the click of a button. Such arrangements are called **recipes**.

Recipes

In a washing machine, there are preferred parameters for some types of clothes. Depending on the fabric to be washed, the operator can select the water temperature, the duration of the wash cycle, and spin dry.

However, for operator convenience, a set of parameters can be predefined (by an expert), for different fabrics. For example, for white cotton, a high water temperature of about 60°C and a long wash cycle is required. If the aim is to disinfect, the water temperature should be 90°C. For delicate fabrics, such as silk, the temperature should be 30°C, and the wash cycle should be really quick, with no spin dry.

Similarly, in an IMM, depending on the variety of plastic being used, and the type of mold that is mounted, a large number of parameters need to be entered. Rather than the operator spending a long time entering all these settings parameters, such parameter sets are stored as recipes. The operator simply chooses the recipe number, and the machine setting is performed automatically.

Events

Events in the machine or plant are recorded along with the time and date for later analysis. These are historic and depending on the storage space available in the controller, they might vary.

Alarms

Alarms are events that indicate some abnormal situation. They serve as alerts to the operator. Alarms require an acknowledgment from the operator when they have taken care of the situation. For instance, if the temperature in a machine is supposed to be a maximum of 50°C and the temperature overshoots to 51°C, then the controller sounds an alarm. It depends on the machine builder if this variation is accepted. If it is, then it only acts as a warning and the machine continues operation. There might be an actual alarm about the temperature being very high if it continues to rise further.

Trends

For analog variables such as level, temperature, and flow, it is useful to observe the behavior of the variable over a period of time. A trend screen can take several formats. The simple variant is that the trend page has preconfigured variables from one to four. The system is always recording these variables. More complex options are possible. *Figure 7.8* shows what the trend of a variable looks like:

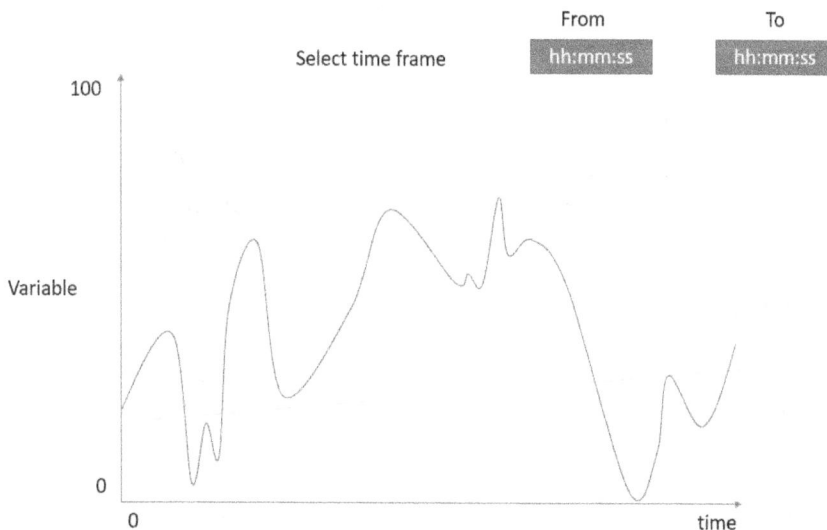

Figure 7.8 – The trend on an HMI

The programmer needs to select the variable and provide a graphical representation of how the variable undergoes a change.

Usually, however, in industry, there are many machines on one shop floor. These machines could all be of a similar type, such as injection molding, blow molding, or FFS machines. There can also be a situation when there are several types of machines in a shop such as a lathe, milling machine, grinding machine, and so on. In this case, there is a requirement for central supervisory control over all machines, to be able to keep track of any glitches or bottlenecks at any station.

Yet another scenario is when machines are operating in a line, performing successive operations on a workpiece. The product of one machine becomes the input for the next machine. Here, there is a need for synchronization or orchestration. The machines work with each other like a rehearsed group of musicians or dancers. This process needs a command center.

We hear data is the new oil. In this section, we took a look into various possible data sources on a machine and how they are represented in the form of warnings, events, or alarms. All data from data sources can be represented in textual or graphical forms, such as trends. We also saw the way parameters can be stored for different parts and objects in the form of recipes. We will now proceed to examine these items, which form a part of the **automation pyramid**.

Supervisory control

Sensors and actuators occupy the lowest level in the automation pyramid. These are part of a machine and multiple machines form a line. However, there needs to be a control system that monitors these individual machines so that the factory has a high-level overview of the various operations. It also makes it easier to send parameters to individual machines from a central place. This is part of the supervisory control. There are various forms of supervisory controls, which are listed in the following sections.

Plantwide information

A typical plant or factory will have several machines. For the efficient running of the operation, some information has to be viewed as an aggregation of each machine's performance. It is also useful to know the present status and alarms from each machine. The location could be a plant control room.

Digital dashboards

There is some information that should be visible from anywhere on the shop floor. This can be production targets, actual production, maintenance schedules, and so on, and can be displayed at a prominent location on a large screen. Many production lines deploy the use of **Andon boards**, which provide the status of machines, lines, and various processes at a single position on the line. These systems can also have audio for alerting supervisors. Thus, it becomes easy for the line manager or supervisor to monitor the production in each shift and track whether everything is on course.

These are typically like the boards seen at an airport. Everyone can stand in front of one board and understand whether the flight is delayed or find out the gate number of the flight.

SCADA systems

Supervisory control and data acquisition (SCADA) systems are an extension of digital dashboards. As is evident, they allow not just viewing data but also the execution of some commands at line or plant level. Usually, SCADA systems are preferred in discrete processes or batch processes.

DCSs

A **distributed control system (DCS)** is a **closed-loop control system (CLCS)** for continuous process plants. A CLCS demands high speed and high safety. Hence, the controllers are not usual PLCs but rather a special electronic construction. The control system also provides special features not available in SCADA. DCSs are used for large power plants, oil refineries, and so on.

Advanced functions

There is no clear distinction today between SCADA and DCS. The main reason is the increase in the capability of PLCs, which can handle closed-loop control as well. Yet tradition lingers in many fields, where the plants are enormously expensive, and a production shutdown for even a short duration can have severe financial implications.

Today, SCADA systems go well beyond monitoring data and executing simple commands. As software becomes more capable, it is not just possible to view the trend of a variable; the system, too, can extrapolate future behavior. It is possible to make what-if sort of simulation studies.

The world is moving to an era of smart factories. A **smart factory** is not only an optimized manufacturing facility but is also automated and digitalized such that it can do the following:

- Enable new product launches depending on market dynamics and needs

- Scale to effectively manage variations of existing products

- Manufacture products and goods in a cost-effective manner

- Have various smart assets, such as machines, sensors, and robots that integrate with IT infrastructure, enabling a high level of automation and seamless vertical and horizontal data exchange

- Have real-time analytics, minimizing downtime and maximizing productivity

In this section, we introduced you to various elements of the automation pyramid that are above the machine-control level. Usually, in a factory, machines from various vendors form a line. These lines interact with either a SCADA or a DCS that forms the supervisory control. Effective operation of all these systems that are fully automated, offering seamless vertical and horizontal connectivity, forms the basis of the smart factory.

Mobile first – control from anywhere

With the **Internet of Things** (**IoT**), we are spoiled for choice. Today, you might be sitting at home and monitoring the status of your car. You can order so many things without visiting an actual store. All of this has added a lot of convenience to our lives and we expect the same in industry. Operators are also demanding intuitive HMIs, just like their smartphones.

The present-day customer wants to interact with machines from anywhere. This means that the operator panels need to be replicated on smartphones so that a subset of operational information can be available, via the web, on the phone at any time, any place. There is also increasing demand that control operations, meaning commands or the setting of parameters, should be possible from anywhere. Systems have the capability to achieve this, but whether it makes sense from the perspective of operations and security is a debate yet to be concluded.

Summary

Prior to this chapter, we have focused on real-world applications and how things get made in real life. However, there is so much more to automation than just machines, lines, factories, and manufacturing. In this chapter, we moved on to exploring aspects of automation, machines, and factories that are not usually discussed. These are essential aspects, especially in today's connected world, but do not get a lot of focus.

Moreover, we considered how much automation is the right amount of automation. There are also doubts over whether automation is essential. Will automation lead to people losing jobs? These are questions that need to be answered. However, autonomous factories at the moment are still far off. Human intervention and logic are needed for machines and lines to function efficiently.

This is a decision that must be taken on a local basis, depending on the situation. However, what is clear is that completely automating the process has drawbacks. More automation is not always better. The best way to frame the issue is cooperative working between people and automation to get the most out of the machine.

"*To complete the example of the washing machine,*" said Josef, "*the best results are achieved when you sort the clothes into cotton, silk, and wool and load the machine by hand, and set the correct program each time.*" Jacob replied, his mood lifted by now, that "*it is too complex to design the automation to sort the clothes and set the program by itself. It is actually simple enough for me to do this part of the task.*"

Jacob had, by now, a good deal of knowledge about how products are manufactured in factories and how things get made. Thanks to Josef, he had good knowledge of the process, the raw materials, and how many everyday products are made. Moreover, he was now able to think about various challenges faced by industry and machines and the possible ways to overcome these challenges. However, this dimension was totally new to Jacob. Josef had opened a completely new avenue in the field of machines and automation.

At the beginning, Jacob had wondered why Josef wanted to explain humans and machines. He thought that if machines are made by humans then they should function independently. However, after an intense discussion with Josef, he realized the importance of humans and machines working together. Additionally, he also realized that as the world moves forward, and new technologies start coming in, the role of such collaboration would be much more important. Collaboration between humans and machines would be of the utmost importance for effective outcomes.

Jacob was now keen to understand what was in store in his next discussion. He now had ambiguous thoughts as his father informed him that they would be discussing the avoidance of human intervention. At the end of this talk, he was convinced that humans and machines need to collaborate and work together. His head started spinning. Josef was quick to recognize this and told him to relax and stay curious, yet not think too much.

8

Automation – Dramatically Helping Avoid Human Intervention

Can we imagine a world without machines? No way! All our daily activities as humans are dependent on machines, either via the machines that are found in factories and shop floors or the products manufactured using these machines, such as our electronic devices, packaged foods, beverages, clothes, automobiles, and so on. In general, most of the other inhabitants of planet Earth—by which I mean plants and animals—adapt to their environment. Humans, via their intellect, have taken a different approach. We adapt our environment to us, for our comfort. The main approach to the modification of our environment is the invention of machines.

There is, however, a flip side to this development. Among other things, tending to machines takes people's time and attention away from other activities. Machines occupy most of our attention today. Just imagine waking up in the morning and not finding your phone next to your pillow. Today, the first thing you might do upon waking up is pick up your phone and see what you have missed during the time that you were asleep. Humans, by nature, are social animals. The invention of machines is meant to make life more comfortable and enjoyable. Tending to machines is not a pleasurable activity. Therefore, machines should be tending to the needs of humans.

We saw various ways that humans interact with different elements of machines in *Chapters 1* to *6*. In *Chapter 7*, we then saw how humans interact with machines at different stages of machine and factory building, commissioning, and machine functioning. Clearly, it is automation that forms the bridge between humans and machines. Automation reduces the level of skill needed to perform a task by handling the micro-steps using instructions previously programmed by an expert. By using this kind of automation, any layperson can perform a task at an expert level.

In this chapter, we will delve into the following topics:

- The need to reduce human intervention
- The control challenge in designing machines without supervision
- Application examples
- How the control challenge is solved

Reduction of human intervention is automation that additionally enhances the skill of the machine and takes it to the level of an expert tool. Machines are often required to run day and night.

The need to reduce human intervention

If we compare the way our parents and grandparents used to start and end their day to the way that we do, we might find a stark difference. This can be witnessed in the difference between how Josef and Jacob do things too. Let us look into their different lives and relate them to how automation and machines have changed lives.

Josef tends to wake up before sunrise. He wakes up, does his morning chores, and then usually goes for a walk or a run. On his way back, he picks up a carton of milk from the store and then the newspaper that the delivery guy has delivered to his doorstep. He spends some time on the couch reading his newspaper along with his morning tea. Josef then gets ready for work before preparing breakfast for both him and Jacob. When Jacob was young, this was also the time Josef would wake up Jacob and try to get him ready for school. Josef was mostly ready long before Jacob was awake. They sat together for breakfast, where they talked about various topics. This was their family time. Josef left for work after breakfast along with his lunch box and returned in the evening after work. As you might have seen, there is no mention of mobile phones or other gadgets, as those were the days when those gadgets were unavailable or only for the rich. When mobile phones were first launched, the providers used to charge around $45 per month for their services, which would include 0 minutes, and then every call would be billed at $0.45 cents per minute. Therefore, a lot of people could not afford such expensive call rates. This was sometime in the early or mid-nineties. In the evening too, after Josef returned, he would freshen up and everyone would sit down for an early dinner, which, again, was family time, and everyone would chitchat. Only after dinner would the family sit in front of the television to watch something.

The following figure shows the changing face of mobile phones over the years. Earlier phones were merely meant for making and receiving calls and for texts. Today, we use our smartphones for everything—shopping, using social media, and even banking:

Figure 8.1 – The changing face of mobile devices over the years

On the contrary, let us see how Jacob begins his day and goes about things today. As he is a college-going teenager, Jacob wakes up because he needs to wake up and start his day. Even after waking up, he stays in bed for some time looking at his phone, glancing at his messages, responding to them, checking his social media accounts, and so on. After he has done all this, he decides to get out of bed. He always has his phone on him as he gets ready for the day. Jacob munches on something before he leaves or simply picks up his bike and bikes to college where he has some snacks along with his friends at the college canteen. He returns home after college and opens his laptop to work on his assignments. After his assignments are done, he switches on his PlayStation to play video games on his television or watches a TV series on a streaming platform. If he is hungry, he opens a delivery app and orders some food, which a person delivers to his house within a stipulated time.

If we compare Josef and Jacobs's day, we can see that Jacob is a digital native and cannot live without his electronic gadgets, whereas Josef is not into digital technology too much and just uses it because he has to and there is no choice.

Thus, machines, electronics, and devices have become indispensable parts of our lives today. Even new homes are now branded as fully automated homes. Today, we try to reduce our intervention in these activities as much as possible. All these examples are from our day-to-day lives. However, things are not too different in industrial contexts. Today, industry is also looking at adding an increased amount of automation to their machines, lines, and factories while trying to reduce human intervention as much as possible.

In general, an automaton is capable of performing a fixed task in a strictly limited environment, the exception being an automaton that's designed to reach a defined stop status and then wait for human intervention. As we have seen, humans are also needed during regular operations. This means that in the production process, humans are employed to perform routine tasks, which require less skill. As part of the progress of human society in terms of development and welfare, it is desirable for humans to be doing more creative tasks. As we progress through the pyramid of Maslow's Hierarchy of Needs, we occupy ourselves with self-actualization. Therefore, for this to be reflected in technological progress, machines should become more and more autonomous and require less continuous assistance from humans.

In this section, we introduced you to how our lives have changed and how electronics, machines, and devices have become part of them. We also highlighted how a similar approach is adopted in industry, with the amount of automation increasing year after year.

The control challenge in designing machines without supervision

Automation is expected to be a sort of stand-in for humans in the case of interaction with machines. Herein lies a contradiction. In *Chapter 7, The Interplay of Humans-Machines-Automation*, we discussed the topic of how humans interact with machines and automation forms a bridge to make this interaction comfortable.

As a recap, these are areas where humans typically interact with machines:

- Monitoring operations while in normal mode
- Design and prototype testing
- Repairs or updates
- Procurement and coordination

Automation plays a role at each stage, by providing a window into the process through **Human Machine Interfaces (HMIs)**, using warning lamps, and so on. A typical tower lamp is shown in *Figure 8.2*:

Figure 8.2 – An automotive line with two tower lamps for machine
running (orange) and a machine error (red)

There are different types of tower lamps—four-colored with buzzers are the most common. This is
shown in Figure 8.3:

Figure 8.3 – A tower lamp with four stacks and a buzzer

Usually, a green light represents machine readiness, an orange light indicates that the machine is running, and a red light suggests that the machine is in an error state. Operation by humans is also enabled by switches and, more sophisticatedly, by operator terminals.

In each case, the human provides a command as a trigger or as an initiator and monitors the progress through an automation device. Automation can give a true picture because there are sensors collecting data, which is relayed to the display device. Similarly, the intentions of the human operator are conveyed to the actuators via switches, relays, motors, and so on.

There are several types of control challenges. In the case of operator commands, the number of commands needed to keep the process going can be pre-programmed. It is easier to train a human operator for certain operations than to build machine automation. For instance, the state of completion when baking a pizza can be judged better by humans using a combination of sight and smell.

In some assembly operations, a component needs to be provided to the assembly arm with a specific orientation. Imagine an operation such as driving a screw for a fixture. The screw must be presented to the fitting arm head-first. It is a question of cost, whether it is affordable to have a complex vision system to sense the orientation of the screw and a six-axis robotic arm to orient it properly and then present it to the fixture, or whether it is more affordable to have a human do the same task.

However, with advances in software, automation can handle most exceptions. These are the requirements for mission-critical projects where a single failure can affect the entire system. Here, concepts such as redundant or hot-standby systems come into play.

Since automation mostly does not involve an awareness of the surroundings, safety issues will make it difficult to reduce human intervention.

Application examples

Let us investigate some more applications in the industry that need human intervention or those that can go on without human intervention.

Continuous processes

There are so many processes that take care of our daily life that must go on continuously. The products or services rendered by these processes are important for us to lead our lives safely. Some examples that come to mind immediately are power generation and distribution, water purification and distribution, and wastewater treatment. These are processes that must occur continuously. Hence, automation must be at the level that a human operator does not need to be continuously present.

Currently, we do not need humans across a waste treatment plant monitoring every single machine. The required parameters can be conveyed to a central control room, and a small number of operators can supervise an entire plant. The control room will have a comfortable atmosphere so that operators are not exposed to an unpleasant environment in the plant.

We spoke in *Chapter 4, Level Control – Controlling the Level of Liquid to Avoid Drying Up or Spilling Over*, of boilers. The feedwater to a boiler should have a low level of minerals. Otherwise, these minerals that are commonly present in the water will leave deposits on the walls of the boiler. These deposits are called scales, and the process is called scaling. Scaling reduces heat transfer and thereby the efficiency of the boiler. If scaling is permitted, the amount of fuel needed to heat the water to the required temperature will increase. The solution for this is a chemical process called demineralization. Some chemicals are added to the feedwater, which will react with the dissolved minerals and convert them into non-soluble solids, which are removed from water by filtration. The amount of demineralizer chemicals to be added has to be calculated based on the amount and type of minerals present in water. The exact correct amount of the demineralizer must be added. With advanced automation, this entire calculation of the amount of the demineralizer (the right dose) can be automated, thus eliminating dependence on humans.

No-second-chance missions

No-second-chance missions are those in which a continuous process cannot be halted, as halting it might lead to wastage, losses, or sometimes restarting these processes, which can be time-consuming. Additionally, as the title of this sub-section suggests, certain critical applications do not provide a second chance to operators, humans, or the entire process. If there is an error or the mission needs to be aborted, there is no going back.

One such example is from the space launch missions carried out by various associations globally. In recent times, besides government institutions, many companies are working toward space launches. Some organizations now promote space travel. As these missions are crucial to add value to human life, there needs to be a huge number of tests and functionality checks before these missions can continue.

A similar example can be seen in the area of autonomous driving. As human lives are at stake, the amount of tests a company carries out is enormous even before they are subject to human tests.

Space launches

Automation and robotics not only play a vital role in spacecraft but also in ground processing, testing, inspection, and other hazardous operations. Each launch needs a massive amount of data from various areas such as testing and command procedures. As you can imagine, these operations are impossible to be done manually by humans, as a single human error might lead to huge losses and damage. Thus, ground automation systems are deployed in these areas to achieve the desired precision and accuracy.

Fueling is another area where automation plays an important role in space missions. In any space launch mission, there is a massive amount of automation and robotics involved, reducing human intervention to a minimum.

A high cost of error

Another set of application areas is those where the cost of errors is too high. These costs could be related to numerous things, whether economic losses, threats to national security, or human safety.

Power plants and rollercoasters are two examples where human error may lead to huge costs, hampering the economic situation as well as human security.

Power plants

A power plant consists of various areas that need automation, such as boilers, turbines, generators, pre-heaters, and transformers, amongst others. There are control systems responsible for controlling each of these individual elements. An effective control system will enable the plant to function properly and avoid any dangerous operating conditions. These control systems deploy setpoint control and various open and closed loop systems.

The power plant consists of electrification, monitoring, control, and measurement, all controlled and supervised by various automated elements. Supervisor control could consist of SCADA or DCS responsible for data acquisitions.

Power plant automation is very complex and critical. However, as you see, there are so many areas of automation, and they all function together rather than in silos, exchanging data with each other. Thus, human intervention in these areas would be time-consuming and, in the case of any human error, could lead to a power outage for an entire area, city, or region.

Moreover, power plant automation is one of the most sophisticated, as a cyber-attack on these kinds of installations can lead to huge blackouts and losses for a nation.

This critical infrastructure proves to be challenging for automation and there is always a need to reduce human intervention to a minimum.

Rollercoasters

We have all ridden on a rollercoaster. It is adventurous and sometimes scary. However, building, commissioning, and operating a rollercoaster is a huge effort. The automation and safety systems need to be well programmed, monitored, and maintained for safe operation.

Moreover, the system needs to be fully automated, with minimal human intervention for faster and more efficient response times. Advanced rollercoasters provide the possibility to run multiple trains on a single track. Thus, interlocks need to be in place in the automated system so that no two trains are on the same block.

Additionally, there need to be preventive maintenance systems in place to check the wear and tear of equipment and avoid any damage owing to faulty components.

This is another example where systems cannot afford to have human intervention and need fast response times to avert any untoward situation.

The following figure shows a typical rollercoaster ride that we see in amusement parks. However, what you may not have realized before is the kind of automation and safety measures behind these joyful and/or scary rides:

Figure 8.4 – A typical rollercoaster

In this section, we introduced several critical applications that demand minimal human intervention and quick response times to avert any unpleasant situation. These applications range from continuous processes to space applications to areas where the cost of errors is high. All these applications, owing to different reasons, are critical and require complex algorithms to be implemented.

How the control challenge is solved

The control challenge is in effect to reduce the need for human operators to be present at a machine or process continuously. There are many ingenious methods invented for this purpose. All these methods make use of intelligent processors, which are smaller, faster, and cheaper than at any other point in the past. With each method, the main objective is for attendance by humans to a problem in production to be made unnecessary, or for attendance to be at a convenient time rather than being immediately necessary. This is what we mean by reducing dramatic intervention.

Stored program controllers

One significant development in controllers is the invention of stored program controllers (PLCs). A PLC has built-in program memory. The innovation comes in the form of a program that can be resident in this memory. The PLC performs tasks as per the instructions in this program. Since memory chips have become smaller and denser, it is now possible to equip PLCs with a large amount of memory. Therefore, it is possible to create elaborate programs that allow the PLC to adapt to various contingencies. In other words, if any exceptional situation occurs, the controller, and therefore the machine, need not wait for human intervention. The various possible decisions that a human would take can now be programmed in advance and stored in the PLC's memory. The PLC program then chooses the appropriate step, thereby reducing the need for human intervention at the time of process operation.

Maintenance techniques

One of our major preoccupations with machines is their maintenance. Maintenance costs, due to the cost of spares, consumables, and effort. Beyond that is the cost of production loss, which is the cost of goods that could not be produced due to the shutdown of the machine for maintenance. On the other hand, it is recognized that well-planned maintenance actually saves costs by avoiding unplanned shutdowns, which are also called breakdowns.

The following figure depicts an overview of the various possible maintenance techniques in industry today:

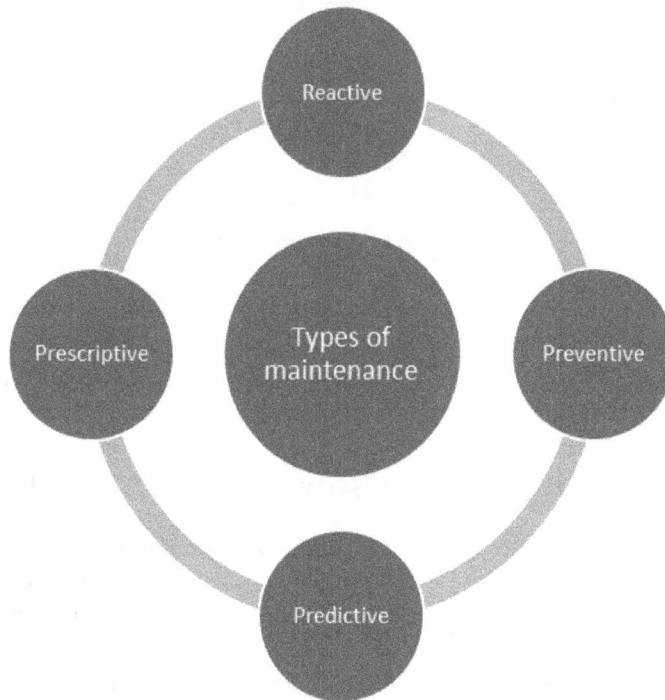

Figure 8.5 – Different types of maintenance in a factory

The following philosophies are in practice for maintenance:

- Breakdown/reactive maintenance: This is also called the if it ain't broke, don't fix it school of thought. In other words, humans keep using the machine until it fails one day. Then, repairs and maintenance are undertaken. The problem with this approach is that the need for repair can come up in very awkward and uncomfortable situations. Hence, prudence dictates going for a better philosophy. Breakdown maintenance needs no help from automation for scheduling.

- Periodic/preventive maintenance: Also known as regular maintenance, this simply means having a regular calendar cycle for maintenance. This can be improved by counting the number of operating hours of the machine instead of going by the calendar date. This reduces the number of incidences of maintenance because if the machine has a long shutdown, there would be no need for maintenance as per the calendar. A further improvement can be to count a weighted measure of hours of operation; the PLC can count the number of hours the machine has worked at full load, the number of hours it has worked at half load, and so on, and thereby count how many equivalent, full-load operating hours have elapsed. The theory behind this is that the wear and tear of machine parts would be proportional to the number of weighted operating hours.

- Predictive maintenance: In addition to keeping track of operating hours, the machine controller can also monitor critical parameters such as bearing temperature, vibration, and so on, and predict the probability of a breakdown the next time. The human operator can keep an eye on this, and when the probability becomes uncomfortably high, schedule a maintenance shutdown.

- Prescriptive maintenance: Here, automation goes with predictive analysis, and instead of just informing the human operator of the probability of a failure, it actually helps out with a prescription for a shutdown. Therefore, here, automation meets humans more than halfway and relieves them of the burden of scheduling.

Systems with redundancy

Redundancy means having more than one resource to solve an issue or task. Especially in continuous process plants, the process must keep running continuously. An abrupt stop in the process or the process running out of control can have disastrous consequences. It can take the entire plant into an unsafe situation—something such as leakage of poisonous gases or an explosion—which can harm life and limb. In these cases, the technique called redundancy is adopted.

In redundant systems, two independent control systems operate on the same equipment in parallel. One is designated as a master and the other is called a standby. The master system controls the process. The standby system reads all the process inputs and also calculates the outputs. However, being on standby, the calculated outputs are not sent out to the process itself. In addition, both the master and standby continuously monitor each other; this is called a health check. If the master fails—that is, the standby system determines that the master is not healthy—control is handed over to the standby, which becomes the new master. There are many smart algorithms and configurations within this topic, which provide increased reliability at different costs.

Fail-safe configurations

Some systems require that, in case of any component failure, the system be taken into a safe situation. An easy example is a passenger lift in buildings. In case of any failure, an emergency brake system operates to bring the lift to a safe place. Then, the passengers can wait for external help. Similar situations can arise in other machines or processes. The system would be designed so that in the case of the failure of a component or of the entire automated subsystem, the mechanism would either turn off or fall back to a state where there is no risk of damage or injury to the machine or any humans.

In the case of moving machinery, going to a dead stop is also not always a safe situation. Therefore, the safe algorithm has to be designed to take care of the optimal status.

Graceful degradation

In a large plant, it must be accepted that at some point, some components will fail. It is not realistic to have a design that anticipates no failures, either electronic or mechanical. However, the intelligent design of a system with thousands of components does not plan for the entire plant to come to a stop in the case of any failure of any component. Graceful degradation means that a failure can be isolated to the affected subsystem, and an automatic reconfiguration of services will permit the plant to continue performing. There could be a degradation in the performance—the plant might run at a lower speed, or only certain products may be manufactured.

Lights-out factory

Lights-out manufacturing is a methodology that has its basis in advanced automation and robotics. Using this foundation, artificial intelligence is deployed with the high-speed collection of big data. This enables the plant to run with nearly no human presence in the factory. If there are no humans at the workplace, then there is no need for lights since the machines will work without a lighted environment – hence, the term lights-out factory.

Before we dive into the benefits of lights-out factory, we will define some terms used in the previous paragraph:

- Advanced automation: This includes functions such as safety monitoring, diagnostics, maintenance, repair, error detection, and recovery. These are next-generation automation concepts.

- Big data: This is the data collected in large volumes from a process continuously and also includes the software and hardware used to archive and retrieve this data.

- Artificial intelligence: This is the term used for automation that is not solely operating on pre-programmed functions but can make decisions based on past experiences as well.

Some high-level benefits of a lights-out factory are as follows. This is only a high-level overview of a lights-out factory:

- Reduced energy consumption

- Reduced workforce

- Reduced inventory

- 24/7 operation (round-the-clock operations)

- Higher productivity

Almost-human machines

As automation progresses to take more and more control over processes, it is time to develop machine concepts that will configure themselves according to production needs.

A couple of years ago, humanoid robots were a widely discussed subject, with Hollywood movies depicting the rise of machines. Today, in industry too, we can see machines are increasingly becoming intelligent. With machine learning, they are learning day by day and adapting to new requirements without human intervention. With collaborative robots, humans and machines are now working hand in hand. Machines are moving in the direction of becoming independent while learning from day-to-day occurrences and results. Artificial intelligence and machine learning, coupled with data and analytics, make the machines sustainable and almost like humans.

In this section, we introduced you to the topic of how various control challenges are overcome and different types of maintenance techniques are deployed in the factories. We also provided an overview of redundant and fail-safe systems together, with the concept of a lights-out factory.

Summary

Today, we cannot imagine a world without devices and electronics. We start and end our day with electronics, wearables, devices, and gadgets. These gadgets provide us with necessary notifications and alarms. Similarly, in industry, automation and control systems provide humans with the necessary warnings, notifications, and alarms. Moreover, automation systems may also take necessary actions to stop processes.

In this chapter, we introduced you to areas of automation where human intervention must be reduced owing to various factors.

Jacob had never realized the comparison of his daily routine with Josef's. He realized that there was such a huge difference in the way they both went about their days. He also realized just how much time he spent in front of the screen and decided to regularly work on a digital detox. Jacob was also wondering whether he could build up a 3D roller coaster that would be made of Lego and controlled by a microprocessor. Josef was impressed and agreed to help Jacob with this project during his semester break.

Now, Jacob was looking forward to the next lecture by his father on how automation could build a super-organism with awareness.

9

Automation Can Build a Super-Organism with Awareness

Jeeves is the ultimate butler. (Jeeves is a fictional character in novels and stories by the famous author P.G. Wodehouse.) The ideal assistant is one who appears when you need something and disappears when the need is served. Jeeves has knowledge about all topics and can answer every question that his master might raise. Jeeves also has inside knowledge about various causes and reasons and can explain why things are the way they are. He has knowledge about relationships between people. While being so knowledgeable, Jeeves is nevertheless unobtrusive and appears only when there is a need. Your need may not be expressed. You don't have to summon Jeeves; he just appears whenever there is a need.

Jacob was wondering whether something like Jeeves could exist in the manufacturing world. He walks up to Josef to check with him if that was possible. After listening to Josef, Jacob was trying to relate automation to Jeeves. I am sure you too at this point in time are thinking about it. Josef indeed clarifies that, in manufacturing too, there is a need for an entity like Jeeves. Such an entity can be provided by automation.

Josef starts explaining to Jacob how automation acts like Jeeves. Jacob was already intrigued and curious to know more. Both sit on the couch and Josef begins his explanation with a coffee in his hand. We are looking at the development of an assistant who is continuously available at the side of every operator, supervisor, and manager in a manufacturing facility. Such an assistant has an infinite fund of knowledge, never tires or forgets, can look into the past and future, and present a hint, a suggestion, an alert, a warning, and occasionally, advice. These elements make up the super-organism. This is how automation can help people to run a manufacturing facility when automation is developed and implemented fully. Such a capability sounds like pure wishful thinking. However, many technological developments are happening independently in different fields. A convergence of these developments could indeed make our wishes come true.

We are looking to expand the role of automation from single controllers dedicated to individual tasks to a group of controllers that performs more than monitoring and control. We ask automation to provide knowledge-based decision support systems. Such developments in automation are ongoing and readily available. We will look at some of them now and also sketch out how they will all work with each other.

In this chapter, we will examine the following:

- A blueprint of automation with awareness
- Technological advances to help realize the blueprint
- Applications where such automation can be deployed
- The challenges ahead

A blueprint of automation with awareness

Jacob was quick to realize that automation is the Jeeves of the manufacturing industry. His assumption was confirmed by Josef. However, Jacob looked lost in thought as Josef was explaining about automation. Josef realized this and asked Jacob whether he had something on his mind or whether he had other things to take care of instead of listening to him that day. Jacob explained that he was neither lost nor thinking about something else. In fact, he was wondering whether automation or the manufacturing industry had a blueprint.

Josef was surprised as well as impressed to hear this. This was a topic Josef was planning on discussing with Jacob that day. Josef enquired as to why Jacob had such a thought. Jacob quickly said that by now he compared automation to an organism having superpowers, like some superheroes. Thus, such an organism must have a strategy or a blueprint. He also mentioned that microcontrollers and chips are rapidly changing. During their previous talk, in *Chapter 8*, he learned about the change in the smartphone market. Jacob did his research and understood that the size of electronics must have changed to adapt to the new technologies. He was not sure whether this was true and wanted to check this with Josef. Josef was again impressed. He thought to himself that half of his work was accomplished, and he could continue discussing automation, as Jacob was on the right track.

Controller chips have been doubling in speed and halving in size every 2 years. The computational prowess of a smartphone controller is more than the command control module of the Apollo 11 moon mission. These small chips get embedded into every device. Each of these controller boards also has the capability to communicate – talk – to other controllers. This is a kind of **distributed intelligence**. A few years ago, intelligence was concentrated in a central controller and processing unit. In a typical architecture, the controller was the main unit for storing all the intelligence and taking decisions based on various inputs. However, today, peripheral devices are becoming intelligent and machines are becoming larger and complex, requiring multiple controllers to control them, which generates a need for intelligent devices across machines, production lines, and factories. Today, such distributed intelligence in multiple devices is very common in advanced machines and design concepts.

Automation becomes more and more a software functionality. The hardware – that is, chips and modules – do not make themselves felt or become prominent to the user or application. Indeed, using the same set of modules, the software can accomplish many different functions. Beyond accomplishing assigned tasks, automation can exhibit other traits too. This is something that we associate with true intelligence – that is, given the same set of triggers, a response is exhibited by the controller, depending on parameters based on surroundings or recent history. This is what we call **awareness**. It is behavior that takes into account past interactions and patterns.

Automation as a super-organism

Automation in the form of controls provides intelligence to devices. Automation can be embedded in each device because of the shrinking size of electronic chips. These micro-intelligent units can communicate with each other using high-speed wireless communication networks. This setup now appears like a colony of small intelligent entities, all of them having different assignments but serving a common goal. The picture that emerges is that of a colony of bees, wasps, or corals. These entities, each small, can act in cohesion to provide great power and intelligence. In a similar manner, small automation units embedded in the different devices in the manufacturing plant can pool together their CPU power and memory to solve computational tasks of great complexity. These computational tasks could be so advanced that they might challenge human capability or ingenuity. It is in this sense that we talk of automation as a super-organism. However, it is clear that such a super-organism does not possess some important traits of humans, such as volition, creativity, or humor.

Automation and awareness

In general, we recognize two types of awareness. One is **self-awareness**. This is a topic of philosophers and has to do with spirituality, which we will not pursue further here. The other type of awareness is **situational awareness**. This awareness indicates a grasp or perception of all the factors about the place and situation. Automation, as we have studied so far, is a unit programmed to perform a specific task and operate at a particular location. In other words, the performance is in a controlled environment. Automation performs a task without taking any input about the environment. However, a more advanced intelligence would benefit by knowing its external environment and might modify its actions based on such knowledge. In particular, if a controller has knowledge (i.e., is aware) that there are dozens or even hundreds of controllers operating *out there*, this can help to leverage this knowledge to increase efficiency and safety.

Let's use an analogy. You would probably picture an orchestra as a large instrumental group that combines instruments from different families. The following figure shows a typical classical orchestra:

Figure 9.1 – A typical orchestra

A similar approach can be pictured in the manufacturing industry. Why is a group of units superior to single units operating in isolation or silos? An orchestra consists of several single players with different musical instruments. Each of them is capable of giving a recital playing solo. This performance has value and can be quite pleasing, but a group of such players, with different instruments playing in coordination, can create a symphony whose beauty is more than the sum of its parts.

Devices with controllers embedded can deliver work output that is of high quality with high efficiency. So, similar to an orchestra, if such devices work with synchronization, then the quality and efficiency can be much higher.

In this section, we highlighted how design concepts have changed from centralized intelligence to distributed intelligence to incorporate various new and advanced technologies. This approach has been deployed owing to the shrinking size of chips and additional processing capacity. Thus, the way forward is to distribute intelligence among the various devices in the machine, rather than concentrating the intelligence in a central unit. With complexities being added with newer technologies, the central processor will need to be more powerful than ever and will in no way be economical going forward.

Technological advances to help realize the blueprint

Technology is rapidly advancing and evolving. It is indeed difficult to adapt to this technological change. When you buy a phone, you might think that you will be up to date for the next few years. Within no time, you see there is a new variant of your phone offering better features. These changes are owing to changing electronics and technologies. There is a similar change in manufacturing, as electronics are the heart of all devices used in this industry.

Controller chips are getting smaller and faster

One of the key developments in this field is the advances in computing processors. There is a law in microelectronics called **Moore's Law**. It is actually not a law but rather a prediction of how **integrated circuit (IC)** manufacturing will develop. About 50 years ago, Gordon Moore posited that computing power in a given size of chip would double every 2 years. Looking back now, the following has come to pass – every 2 years, the size of the chips halves and the speed of computation doubles. This means that over a period of time, microcontrollers have become so small and so powerful that it is possible to embed them in every device – in every switch, valve, meter, and so on.

Internet of things – controllers can talk to each other

Since controller chips are getting smaller and more intelligent, it is also possible to endow them with the power to access the World Wide Web using wireless technologies. The significance of this is that it is not only possible to embed intelligence in every device – be it a gadget, tool, sensor, or actuator, all of which become intelligent as a result – but, beyond that, these devices can also access the internet. This means they can communicate with humans all the time. Beyond that, the devices can actually talk to each other, meaning that they can share information. This sharing of information by devices can be autonomous, meaning it need not be triggered by a command from a human.

The following diagram shows how wearables are changing how we do things:

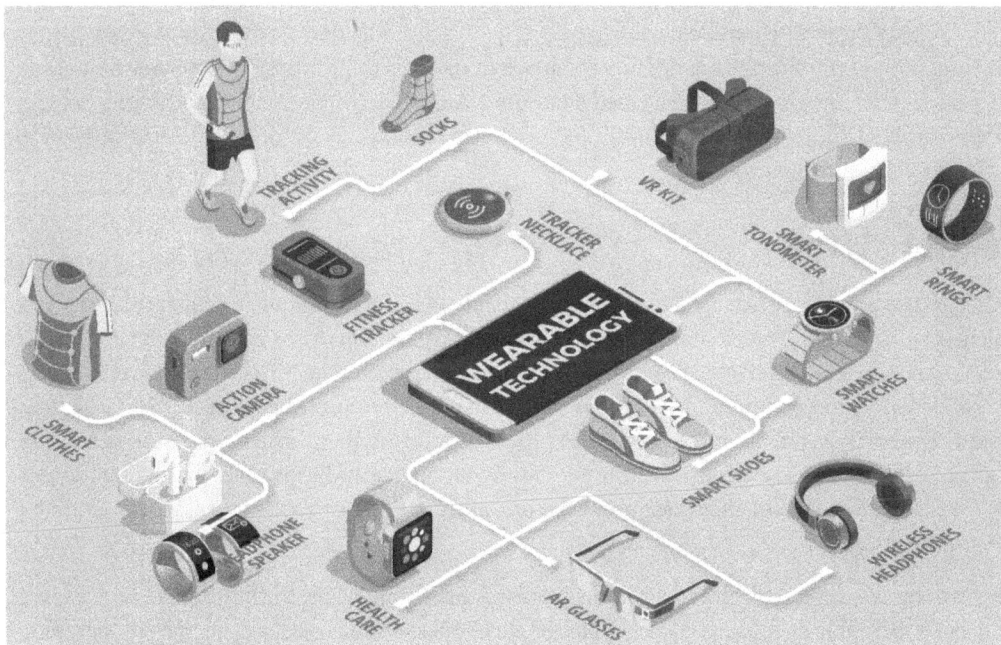

Figure 9.2 – Wearables becoming part of our daily lives

Just imagine wearables that can connect to the internet and track various parameters. It is simply wonderful how a watch can track so many things. Today, you can make calls via your smartwatch too.

Pervasive computing

Pervasive computing helps people interact with information easily by unobtrusively embedding computational power into an environment. This enables users to integrate non-intelligent devices into a smart environment. This means that information is available at any point of time in a factory, no matter from which aggregate the data was originally collected. This points to storage at a central place – namely, the cloud. Retrieval of the information will use algorithms to evaluate the relevance. The information is not just the raw data as it was collected, but data that has been processed and evaluated before being presented. You can now see how **pervasive computing**, used together with **augmented reality** (**AR**), can form a Jeeves-like entity.

Pervasive computing finds application in many fields – in the tracking and tracing of people and products, finance applications, and medical care. However, significantly, it forms a key ingredient in smart homes, smart factories, and smart cities.

Cloud and infinite processing power

The cloud enhances the capabilities of the billions of devices that are connected to the internet of things. Cloud computing is a strategy by which data storage and CPU power can be used without actually owning these CPUs and data storage devices. It is a model that works on a pay-as-you-use basis. This has big implications for devices that are hooked to the **Industrial Internet of Things** (**IIoT**). Each device might only have limited computational power and a small database and storage, but they are able to access the nearly unlimited power and storage facilities in the cloud and so are capable of a very large intelligence.

Augmented reality

The data that is acquired through sensory input is reality. AR is a technique in which this reality is supplemented with more information obtained from cyberspace. This information from the cloud is relevant to the context. This context is in three dimensions: space, time, and process. The space context is the particular aggregate under observation—which machine or unit is being dealt with? The time context is the reaction time that is desired. The process context is about which operation is actually being carried out – is it regular production, or is it maintenance and repair? The augmentation of reality should be helpful and relevant to these factors; otherwise, it actually works as a distraction by cluttering things up.

Reality can be augmented by either audio mechanisms or visual effects. Both methods can operate by the use of wearable gadgets, such as goggles or earphones, or by projection, such as holographic effects. Further development could be bionic devices or implants, which we will discuss in *Chapter 10*.

Data analytics

Data analytics refers to computer algorithms that look for and find patterns in data. As an extension of IIoT, this is a very powerful tool. Devices that are connected to the internet communicate data about their activities to the cloud, where it gets stored continuously in huge repositories called **big data**. Analytics engines can pore over this data and find trends or patterns to produce insights and generate predictions. These are useful for humans to make more efficient decisions. They are also useful for devices to present a view as a cognizant and aware organism. Indeed, an organism that has more computational power, memory, and data than many human beings can be called a super-organism!

Pattern recognition

Recognizing patterns is a defining feature of intelligence. Pattern recognition helps us in many ways. Recognizing a pattern in a series of events enables us to predict possible outcomes. This in turn enables us to take preventive or corrective action at an early stage. When a machine is monitored continuously, a huge amount of data is generated. Such data simply examined as a time sequence may not yield much insight. It is surely useful to monitor parameters within safe limits of operation, but to find out the cause of deviations, you have to derive patterns of behavior. Pattern recognition consists of three steps: representing data in a suitable form, recognizing features that are meaningful for the task at hand, and classifying the feature set.

Pattern recognition finds use in many activities of manufacturing. Most frequently, it is made use of in maintenance scheduling. We discussed in the previous chapter (*Chapter 8*) the different philosophies of maintenance. All of these base themselves on **artificial intelligence** (**AI**) and particularly on the capability of pattern recognition.

Beyond the field of maintenance, pattern recognition plays a useful role in other areas – for instance, computer vision and image processing, speech recognition, and face recognition.

Classification

A frequently performed task is classifying products and situations. In the case of products, it could be to classify an object as good versus no-good in a quality inspection. In the case of situations, it could be classified as safe versus unsafe. The key point to note here is that it is often not easy to define the parameters by which this classification should be made. It can actually be a combination of a set of parameters. In such cases, we make use of **Machine Learning** (**ML**).

Machine learning

In situations where it is not possible to define exactly which parameters or limits yield a meaningful classification, ML is used. ML is a branch of AI. ML actually derives the rules for classification by examining a large set of data samples.

In the programming of automation systems, the automation code generally consists of a set of rules called an algorithm. With the algorithm, the automation unit monitors the input parameters and provides a signal to the control unit to bring the process parameters to the set point. As an extension, an AI unit is provided with an algorithm – that is, a set of rules – and, by evaluating the inputs, provides guidance or prescriptions to enable better decisions.

ML works in a different fashion. Here, the automation is provided with a large number of examples, each of which is labeled. This is called a **labeled dataset**. The machine examines all these samples and extracts the common features of those samples belonging to a particular class. For example, consider the problem of examining the surface finish of a painted vehicle body. The machine, after examining many samples of a perfect finish and many samples of a blemished finish, can then set forth and classify further samples as perfect (i.e., passable) or imperfect (i.e., rejected). Note that the process must also work with respect to the various colors and shades that are used.

The features this machine uses to determine what is good and bad are actually not accessible to the human operator, nor is it always possible to understand why a product is rejected. This is because the rules are not taught by a human to the machine; rather, the machine has learned by itself after a series of experiences. This is the way a human learns many topics too. Some things, such as multiplication tables, are taught. This is how the large body of AI works. Some things are learned, such as slowing down because you are approaching an unsafe situation while driving. This is learned through past experience, and sometimes, we are unable to explain why we classified a situation as unsafe. This is close to how ML works too.

Robotics

When we talk about an assistant to a human operator, the picture that springs to our mind is that of a robot. The picture may be generated from some science-fiction movies. Such robots are depicted as humanoids trying to imitate humans. However, such depictions in these movies sometimes easily fail, where we as an audience can distinguish them from humans. In technology, it is unwise to write off anything as pure imaginative fiction. Such imagination or fiction is increasingly becoming fact within the course of a generation.

As we touched upon in *Chapter 7*, robots are a common presence in smart factories. By deploying robots to perform some tasks, we can reduce human intervention in the manufacturing process. As technology develops and as smaller processors and built-in communication becomes more available, we can set higher expectations from robots. Some areas where we can expect to see rapid progress are as follows:

- **Mobility or locomotion**: To monitor or take actions, it is necessary for a robot to be capable of moving around a factory. We call this **locomotion**. The next step would be to integrate a robot and a drone, which allows the locomotion to become even more free-range. The challenges here are planning a path for travel, recognizing obstructions in the path, and autonomous modification of speed and trajectory with awareness.

- **Sensing and perception**: Mobile robots need to be equipped with sensors that go beyond monitoring the workpiece (product in manufacturing). Robots need to be equipped – for example, with vision – so that they can detect obstructions in the path of movement. They also need to recognize objects and their orientation. A seemingly simple task such as placing an object on a table can present great difficulties. The robot needs to check whether the surface is flat, whether there is free space, and so on, which are intuitively obvious to a human being.

In this section, we provided an overview of technologies that are upcoming or being used in the industrial automation and manufacturing industries, how the IoT in the industry is changing the manufacturing landscape, and how manufacturing has evolved.

Applications where such automation can be deployed

You might now be interested to learn about various applications being built using these technologies. In our day-to-day lives too, we hear about various technologies, but we might not know how they apply to actual applications. In this section, we will introduce you to various applications and how technology is being adopted to make things easier.

Quality control

Quality control refers to producing goods or services conforming to specifications. These specifications are largely claims set out by the manufacturer, who will list a set of features that their product will provide. There are also specifications set by regulatory authorities who, in general, wish to protect the interests of consumers and also their safety. In order to provide high-quality products, it is necessary to have high-quality inputs and raw materials. Hence, a raw material inspection is a major task for manufacturers. Another source of poor quality is the process itself. If the process introduces deviations, then, of course, the final product will not meet specifications.

Wisdom suggests that recognizing deviation at an early stage is beneficial. If there is a part with bad quality, it will be rather inefficient to propel it through all further stages of processing only to be rejected at the final stage. Extending this logic further, it would be prudent to include the supply chain, the vendors of input components, in the quality management system. This resonates with our discussion in the *A blueprint of automation with awareness* section.

Supply and logistics coordination

In a manufacturing process, many entities are linked to each other as suppliers of components. Each unit (machines on a factory production line) performs some processing on the product and moves it up the chain. In such cases, close monitoring of the production requirements and availability at each stage is needed to keep the entire operation working smoothly and efficiently.

The following diagram shows a typical high-level overview of the supply and logistics in a milk factory:

Figure 9.3 – A typical milk production flow – supply chain in milk production

Thus, to manufacture milk packets, the factory needs to ensure the availability of raw materials. Thus, milk is the primary raw material. However, they need to ensure the raw material is delivered to the factory in time in a temperature-controlled environment; otherwise, there are chances of the milk spoiling. Thus, time is critical here. Milk can be packed in pouches or cartons. The factory needs to ensure the availability of these raw materials, their lamination process, filling mechanisms, labeling machines, the milk's packaging, and transport to its destination in a temperature-controlled environment. Thus, supply, logistics, and time are critical in this setup.

Safety

Safety is everybody's business and also nobody's business. As a quip, it may be funny, but safety is no joking matter. One very important aspect of safety is to recognize early on an unsafe situation. This is a not just a static condition, and we have to adapt to changing circumstances – for example, a narrow aisle through which a forklift is traveling and a person is about to step into. Unsafe situations can develop very quickly, and to avoid them, monitoring many factors simultaneously and recognizing a pattern as unsafe is needed.

Smart factory

A **smart factory** obtains efficient performance by using interconnected machines, communication networks between machines and distributed intelligence. The idea is that by having a collaborative operation among machines, production delays and build-up of inventory of raw materials, finished goods, and intermediate products can be avoided. Each machine or process can be optimized by developing good control schemes, but greater efficiency is obtained by observing a factory as a whole. Indeed, it is expected that going beyond the factory and integrating an entire supply chain can have a much more beneficial effect.

The following diagram shows a high-level overview of smart manufacturing. These are some elements of a smart factory or a factory of the future:

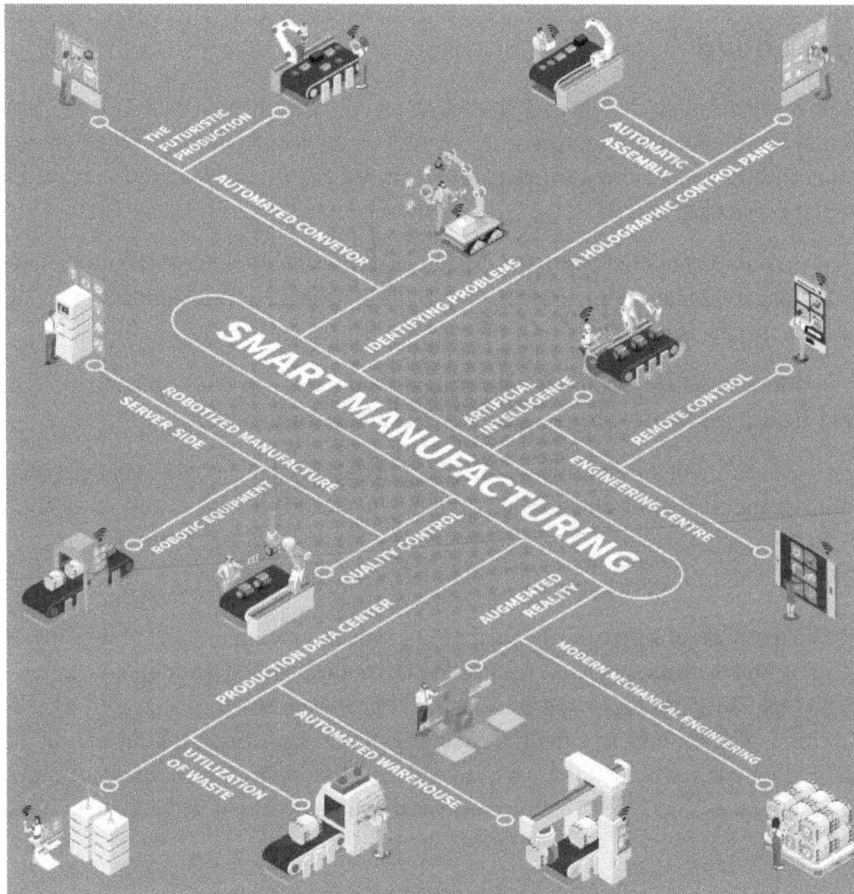

Figure 9.4 – Elements of a smart factory

Factories are made smart by deploying pervasive computing.

Autonomous vehicles

Driving a vehicle through traffic or terrain is a very complex skill. The challenges involve keeping control of the vehicle, planning the route, interacting with other traffic, ensuring the safety of others and yourself, and dynamically adjusting to the immediate environment as well as taking into account data about the environment further ahead. As we have been moving through this chapter, we have learned about many developments in the field of automation, which addresses similar challenges. The main issue here is that all of those challenges are presented in a vehicle that is moving at some speed and where the possibility of manual override is small. In addition, there is the ever-present risk of harm to life, limb, and property. Just as developments in automation have aided the design of autonomous vehicles, equally, the development of driverless vehicles has promoted development in automation.

Challenges ahead

Most of the wonderful solutions that we have learned about are dependent on data. This data is needed in ample volume and in real time. It emanates from various units and entities. Collecting such data and storing it costs effort and money, which gives it commercial value. However, to make the best use of the data, it needs to be shared.

This sharing of data among related entities is called collaboration. Looking at the mutual benefits, related organizations might be willing to share data, but if it belongs to competitors, it may be more difficult to reach a sharing agreement.

Beyond competition, there are also issues of ownership of data, privacy, and security. These are challenges that are being addressed by regulatory bodies and governments.

In this section, we have highlighted various applications of advanced technologies in manufacturing. We also focused on the challenges oragnizations will face in the future.

Summary

Jacob now had an overview of where the manufacturing and automation industry is heading. He was now in a better position to understand how automation forms the building block for new technologies, such as Industry 4.0, IIoT, and digitalization. Jacob was now even more curious to read and learn about these technologies. He now had an understanding of the engineering subject he was studying that, through his discussion with Josef, was coupled with real-life applications of these concepts. With today's discussion, he was now able to envision where the manufacturing and automation industry was moving

Thus, he now had an upper hand over his classmates and peers. He was also able to comprehend complex engineering principles, as he was able to relate them to what he had learned from Josef.

Jacob was not at all exhausted from this discussion, as his thought process was already going in the right direction. Moreover, Josef too was pleased that his discussions with Jacob were helping him progress and enhancing his logical reasoning capability.

Jacob wanted to continue his conversation with Josef, but Josef had some office work and informed Jacob that he would continue this discussion later.

Jacob was unhappy, but he accepted the situation and then went to his room to work on his do-it-yourself project, trying to see whether new technologies would improve his project. Jacob was now eagerly waiting for a discussion with Josef on what's next in the fields of manufacturing and automation.

10
What's Next?

Humans have unique capabilities that automation can never achieve. Hence, there is no place for fear that machines will rule the world. It is not simply a matter of superiority. Automation will eventually be superior to human capabilities. However, there are traits that are unique to humans that set us apart.

In this chapter, we will look at the following topics:

- The question
- Need for the question
- What are the challenges?
- How could automation address the challenges?
- Solving human problems

Additionally, we will provide guidance on what's in store in the future with automation.

The question

The contribution of automation to the overall well-being of humans is undeniable. In the evolution of humans, there have been several important steps. The last one hundred years have been dominated by automation technologies. Today, we can hardly find a product, service, or process where automation does not aid in obtaining better results. Our quality of life is improved by automation helping out in producing these articles or services of daily use. Automation reduces human effort in manufacturing; it also provides better control over the process, thereby increasing throughput and quality. Now it's time to ask, "What's next?"

There is a question that every young aspirant dreads at a recruitment interview. The interviewer asks so casually, *"Where do you see yourself in the next five years?"* In the same vein, *"Where should we envision automation in the next two decades?"* The question has to be answered both in terms of automation technologies and also in terms of the applications to which automation will be put. Crucially, the answer will also contain elements of how the interaction between humans and automation will develop. Another very important point to look at is the evolution of automation over a two-decade period. We see technology evolving rapidly. However, these technological changes and advancements are much faster in the consumer space or the **Internet of Things (IoT)**. In the automation area, these changes are rather slow. The changes seen in automation a decade ago were even slower. Thus, the speed of adoption is rising but it is still slow compared to the consumer space.

Josef ask Jacob when he procured admission to an engineering college, *"What do you plan to do after your engineering course is complete?"* At the time, Jacob had no clue what he would do. He took a few days, asked a few friends, and then came back to Josef to tell him he might consider doing his master's and moving to the US. Today, after lecturing Jacob for the past few weeks, Josef thought it would be a good time to ask Jacob the same question. He asked Jacob, *"Now that you have completed two years of engineering and after our interactions over the past few weeks, what do you plan on doing after you finish the course?"* Jacob responded instantly, *"I would like to program automation systems and build faster, more efficient, and highly productive machines. Yes. I wish to be a machine programmer."* Josef was pleased to hear his answer. However, Jacob had a question for Josef. Jacob asked Josef, *"Technology is evolving rapidly. Humans are developing these technologies in some corner of the world and then others are following the trend. It is exactly like a video going viral on social media, where someone starts a trend and then others simply follow it. Making a video is relatively simple. However, if a machine is built and then after some months there is a new technology that is more promising, then how would someone use it? Would the machine builder provide it free of charge, or would there be charges? How easy would it be to upgrade these systems?"*

Josef responded, *"This is exactly like your smartphone. Today, you buy a phone, and you can do everything on it. However, you buy a phone and the following month the next version is launched. What would you do as a client and what would the smartphone provider do as a manufacturer? Would you throw away your smartphone and incur additional costs for a new phone or would the manufacturer upgrade it free of charge?"* Jacob said, *"It's impractical to spend such a large amount of money again within a month or even a few months. Also, the manufacturer would have to incur huge costs to replace these old phones"*. Josef said, *"This answers your questions. It is impractical to incur such high costs. However, if it is purely a software upgrade, you would consider doing it and it usually happens at the click of a button. This is similar in manufacturing and automation. It is difficult for factories to replace hardware with new hardware. Moreover, factories have used automation hardware controllers and software for decades. They do not upgrade it as often as in the consumer domain."*

In this section, we raised an important question that every student dreads, which sets the tone for this chapter. Automation and manufacturing are typically fields that not all are aware of, as engineering disciplines do not provide easy guidance on entering this area. Moreover, professors are also not aware of this field and the activities within it and thus are unable to fully guide the students. Automation and manufacturing have thus far been areas that have not been very lucrative or attractive as they were never in the limelight. Moreover, companies too did not interact with universities, making it difficult even for professors and students. This was due to the fact that the workforce requirements of these companies were not very high. Today, the scenario is changing; in automation, manufacturing companies require more people to meet their customers' needs. In corporate life, *"What's next?"* is a question one might face either in each interview or in their yearly appraisals.

Need for the question

Once we have raised awareness of the most important question, we need to look into the need for such a question.

Can we simply extrapolate?

As we examine the progress in the field of automation over the past decades, we do recognize relentless growth. On one hand, we see the power of processor chips increasing all the time, and on the other hand, the size of the chips is reducing all the time. Communication technology is improving too, and speeds of data transmission are improving by orders of magnitude. 5G has already been introduced in the IoT and consumer space, taking connectivity and data exchange to a whole new level. It is now making its way into the manufacturing domain. Thus, we are on the brink of 5G communications in automation and manufacturing domains, which we all know is an order of magnitude faster than the earlier generation. More intelligent algorithms for data compression would increase speed all the more.

Hence, if we plan a system for the next decade, should we envisage the existence of controller CPUs with ten times today's capability? Will the shape and form of automation remain the same as today, or will it be different? This is why the question of *What's next?* is relevant.

What can put a limit on the growth of automation? For applications of automation, human ingenuity is the only limit. As we know, there is no limit to human ingenuity. Only theoretical limits apply to the power of the processors and communications. As clock speeds (the clock/quartz crystal in processors and controllers) increase, we may approach the speed of light that can enable transistors to fire quickly, increasing the speeds of operation further. But we are nowhere near this limit.

Is it too much of a good thing?

It is no doubt that automation has had beneficial effects on mankind. But like all technology, there are also some downsides to it. Automation is used also for making products, all of which are not good for our well-being. A major effect is of course the bad effect on our environment. One effect is that production processes (which all use automation) produce waste products as well, which endanger the environment. Beyond that, by promoting the consumption of more and more products, we are also running the risk of exhausting non-renewable natural resources. The blame for all this cannot be laid at the door of automation. Automation is a tool, and mankind should go about deploying it in a responsible manner. But the availability of such a tool with a potential for misuse does raise qualms.

There is also an apprehension that by getting mankind addicted to automation, we become lazy. We lose the capacity to enjoy manual work and the imperfect products that come out of it. This argument can be extended into the domain of philosophy and the fundamentals of morality.

When do we call a stop to it?

So, is there a time when we say, *"That's enough"*? Actually, progress, particularly of technology, has an inherent force of its own. As long as humans possess a sense of curiosity, there will be exploration. Out of this exploration will arise new insights, new technologies, new methods, and new products. In as much as curiosity is a defining trait of humans, progress will continue. Automation will continue to become more intelligent, more powerful. There will be an unending series of new applications where automation is deployed to our advantage.

In this section, we explored the situation in the chip market and how controllers and processors are changing year-on-year. In parallel, there are breathtaking developments in the technology of communications, specifically in communication between machines. The effect of these changes in the chip technology affects how automation and manufacturing landscapes will further change.

What are the challenges?

What will be the features of the automation of tomorrow? How will it be different from today? In other words, what should we, as automation engineers, work on today?

Improving functional capability

When we look down the path ahead, we can quickly see that improving the functional capabilities of automation is a challenge. This challenge mainly resides in the capabilities of sensors and actuators. If we want to think of topics such as **pervasive computing**, we need to have sensors that are much smaller and that have lower demands on power. In fact, we should have sensors that draw their power from the object that is being monitored. But still, in today's view, the functional capabilities of sensors or actuators are not the biggest issue.

Improving interaction

No matter how advanced it is, automation needs to have an interface with humans. We saw in *Chapter 7, The Interplay of Humans-Machines-Automation*, that automation needs human intervention at different stages. Currently, this interaction is tilted toward automation; information is formatted to be relatively easy for automation to consume, and it takes human effort to format the information in this way. For example, we prepare instructions for machines using languages that are very different from the natural languages of humans. This method has, of course, generated huge employment for programmers, but surely it is not the most efficient manner in which to go about this job.

Which functionalities are to be avoided?

There is underlying disquiet about the growing capabilities of automation. Many people are apprehensive that **automation** will take over the world. This panic mindset has existed for a couple of hundred years, mostly promulgated through fiction. So really, the functions that we develop should take care not to stoke such fear, nor make it more widespread. Today, machine learning and artificial intelligence are being increasingly used in machine automation. These are coupled with automated guided vehicles, **collaborated robots** (**cobots**), and the like. Thus, combining machine learning and artificial intelligence with such robots is the way forward. There are major companies that have a deep focus on keeping track of their automation intelligence, especially in the IT space. In years to come, such departments and domains might open up in the automation industry too, so much so that philosophers and ethicists will work on monitoring the amount of automation and its effects.

Improving security

Security was never a primary focus for automation systems. However, in the past decade, there has been an exponential rise in cyberattacks on manufacturing plants and automation systems. Since 2010, we have seen a constant rise in cyberattacks on manufacturing setups. These attacks have intensified since 2020. **Ransomware** is one of the most common cyberattacks on manufacturing setups witnessed recently. Ransomware is where hackers gain access to critical data and lock it. The hackers then demand a ransom for unlocking the data against a guarantee of not leaking it on the internet. Hackers have understood that these manufacturing setups are soft targets and there are many loopholes. A few years ago, automation experienced a similar challenge with safety. At that point, security was a good-to-have feature but not a must-have. However, today, security is of primary importance for factories and automation vendors, and it is no longer just an option but a necessity. Thus, control systems are needed to not only provide the necessary control and algorithms but also security against internal and external threats.

In this section, we highlighted some challenges that originate from more automation and pointers to keep in mind when the amount of automation increases. We focused on collaboration between machines and humans, adding functionalities in automation devices, and improving the security of devices to protect them from hackers.

How could automation address the challenges?

Without doubt, these challenges can be overcome by automation and control systems. It goes without saying that new-age control systems are needed to overcome these challenges and legacy systems would eventually need to be changed.

Quantum computing

There are many new technologies on the horizon that promise to deliver higher processing speeds with smaller chip sizes. There is a theoretical limit to how small chips can become, but there are ingenious workarounds for that problem. One promising line of development harnesses the power of quantum electronics to perform calculations at high speeds. However, these developments first aim to improve processors for large calculating machines, such as supercomputers.

As far as microprocessors and microcontrollers are concerned, the existing technologies can continue to deliver improvements for another decade or more. This means if Moore's law continues to operate, and we find no reason for it not to, using the same processes of optimization, we can achieve another order of magnitude in the processors for **Programmable Logic Controller (PLCs)**.

Low power consumption

One area where improvements are sought and will be achieved is low power consumption. With the decreased power consumption of processors, the deployment of PLCs can be off the main power supply. This has great implications for applications that need processors to work in remote locations, on mobile devices, on handheld devices, and so on. Low power consumption also means the need to charge the battery is infrequent.

Context awareness

In advanced automation, we expect the automation controllers to be aware of the context of the activity. This, in general, means the awareness should be around the field of deployment, the prevailing physical environmental parameters, and the possibility to extrapolate to the near future. This can happen because no controller will be working alone but will have access to the cloud.

Unobtrusive functioning

We discussed in *Chapter 9* the unobtrusive Jeeves, who only appears when needed and then quietly disappears. This is the capability that automation will acquire when it doesn't make itself felt when the situation does not call for it. If the process is running within design parameters, there should be no unnecessary popups or alarm bells. If a status check is desired at all, only then a status diagram or a dashboard would appear and can quickly be dismissed when it's no longer needed. This helps the operator or supervisor to concentrate on other tasks.

More modes of command input: vocal, visual, mental

The present modes of giving commands to controllers are by pressing buttons or keys or touch hotspots. But to be able to do that, the corresponding input device needs to be present. As the next step, inputs by voice commands could be possible. Indeed, already we have automation with voice recognition capabilities, such as Alexa. Beyond voice, however, automation can follow our gaze; when we look at a particular operational field, the controller will perform the task as indicated by the movements of our eyes. But the ultimate advancement would be if the automation could actually follow our thoughts and get activated by thinking about a task. This can be done today at an experimental level. However, tapping into the brain waves is done by placing electrodes on the body. This, too, we hope will become redundant in years to come.

Updates on the fly

As of today, and also in the future, every automation controller needs updates. As more and more functionality moves to software, the updates will be mostly for software. At present, the procedure to implement an update is to shut down the controller. Sometimes, updates are done online when the controller is active, but for the updates to take effect, the controller needs a power cycle. This is identical to our laptops: once the system updates, it requests us to restart so that the updates can take effect. Next, you have to establish a connection to the controller and erase the old software. Then the updated software is downloaded and the system is restarted. But, in the future, we can expect that while the controller is running its tasks, new software can be downloaded, and once the download is complete, the new software gets activated. This means that this necessary housekeeping activity happens totally unobtrusively.

Security

Control systems and networks are now being built using the **secure-by-design principle**. Moreover, there are standardization organizations such as the **International Electrotechnical Commission (IEC)** and **International Organization for Standardization (ISO)** that are standardizing security in controls and networks at the **operational technology (OT)** as well as **information technology (IT)** levels. Security is looked at as a productivity killer. However, with the rise of cyber threats, it is essential to find the right balance between productivity and security. Additionally, security adds to costs, and yet again it is up to the factory, machine builder, and automation vendor to strike the right balance.

Solving human problems

At a broad level, we look to automation to solve human problems. A couple of centuries ago, the problem that was urgent was to substitute muscle power with other means for driving machines, vehicles, and so on. The answer was to control the power of steam to provide motive force to drive ships, locomotives, spinning machines, and a variety of such mechanisms.

All animals evolve. They change to suit the environment, the ecosystem that nature provides. Humans have taken a different path, mainly because they are endowed with intellect. They attempt to surround themselves with an environment of their own making, according to their own comfort. However, the devices and gadgets engineered for this purpose have, over a period of time, turned hostile to Mother Nature. As a result of this behavior, some problems that have arisen are pollution, climate change, and, arising from climate change, violent natural disasters. These problems need urgent attention.

The problems have given rise to a VUCA mindset. **VUCA** stands for an environment with **volatility, uncertainty, complexity, and ambiguity**. In short, there is a pervading atmosphere of anxiety that is unpleasant.

This raises a few questions: *is this development actually sustainable? Will it yield good results for a few years, a few decades, and then plunge mankind back into a worse condition than where we set off?* That must be avoided at all costs. So, our progress henceforth must have sustainability entrenched and must honor and protect nature. That is how we can come out of the VUCA attitude.

The central point is: will automation stay like Jeeves, as a servant, or will it grow to become a genie out of the bottle with a mind of its own, capable of wreaking mischief and mayhem?

There is yet another factor that is also important. In our race to use automation to provide a more comfortable existence, the benefits have not been distributed equitably. There is more inequality today in the world than at the beginning of the first Industrial Revolution. A major battle to be now fought is against this inequality.

In all of these battles, we will have our faithful companion automation helping us. So, when we move forward, automation too will move forward. We should not have apprehensions that automation will take over the world. Rather, we should view all-pervasive automation as Jeeves, who knows everything and is always ready to come to our assistance. Just as Jeeves is always there to help out in any activity, automation is always available to humans to support them with tedious and complex tasks. Automation and humans coexist today and will continue to coexist in the future.

Summary

We are sure this book was filled with real-life applications regarding how automation helps humans to manufacture products that we use daily. We saw how humans have built machines, lines, and factories for manufacturing various products. In Jacob, we saw an engineer who will witness how the world has moved forward and how they can apply various principles they study in college.

As we witnessed, the world of automation is vast and exciting. Jacob was exposed to the world of automation during his internship, and it was an eye-opener for him. With Josef's help, guidance, and mentoring, he was able to build on the principles and concepts he was studying in college. Over the entire exercise, he was able to build various small kits and program them for usage in his daily work. Some kits included drip irrigation for his small garden, water level management of the overhead tank at his home, and some small home automation projects.

Toward the end of the book, in *Chapters 7 - 10*, we explored technology and its advancements; how technology is evolving rapidly and how manufacturing units can keep pace with such technology changes.

It was a Sunday afternoon; Jacob and Josef were sitting for their afternoon tea and coffee in their garden. Josef was impressed with Jacobs's projects and their effective use. Josef was unaware that Jacob was able to implement all these projects without any support. Jacob had already realized that the journey of automation was coming to an end. Jacob was now more confident than ever. What differentiated him from his peers was the practical knowledge and experience he got from Josef's talks and work. He was now able to envision various applications and engineering concepts and their usage in the industry. He also was now able to understand various interdependencies of different engineering streams, such as electrical, mechanical, and electronic. Additionally, Jacob was also aware of the need for these different departments in industry.

Josef was thinking about what he could explain today and that it was now time for Jacob to experience everything on his own as the new technology advancements would be alien to Josef. However, before Josef could start, Jacob spoke up, *"Thank you for the wonderful experience, enabling me to think about how concepts translate to actual applications. It has indeed been an exciting journey. Thank you for equipping me with this immense knowledge that is not captured in any of the syllabus and college books. Even reference books do not bridge this gap and offer such vast knowledge in a single place. I know it is now time for me to start experiencing everything on my own. Your guidance has made my life much easier, and I am sure this strong foundation will help me in the future."*

The future is exciting. Humans and technology will work together hand in hand, enabling a sustainable future.

Index

‹packt›

Packt.com

Subscribe to our online digital library for full access to over 7,000 books and videos, as well as industry leading tools to help you plan your personal development and advance your career. For more information, please visit our website.

Why subscribe?

- Spend less time learning and more time coding with practical eBooks and Videos from over 4,000 industry professionals

- Improve your learning with Skill Plans built especially for you

- Get a free eBook or video every month

- Fully searchable for easy access to vital information

- Copy and paste, print, and bookmark content

Did you know that Packt offers eBook versions of every book published, with PDF and ePub files available? You can upgrade to the eBook version at packt.com and as a print book customer, you are entitled to a discount on the eBook copy. Get in touch with us at customercare@packtpub.com for more details.

At www.packt.com, you can also read a collection of free technical articles, sign up for a range of free newsletters, and receive exclusive discounts and offers on Packt books and eBooks.

Other Books You May Enjoy

If you enjoyed this book, you may be interested in these other books by Packt:

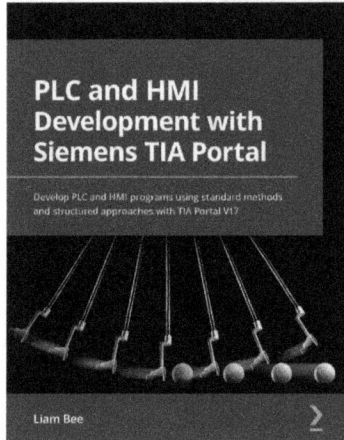

PLC and HMI Development with Siemens TIA Portal

Liam Bee

ISBN: 9781801817226

- Set up a Siemens Environment with TIA Portal
- Find out how to structure a project
- Carry out the simulation of a project, enhancing this further with structure
- Develop HMI screens that interact with PLC data
- Make the best use of all available languages
- Leverage TIA Portal's tools to manage the deployment and modification of projects

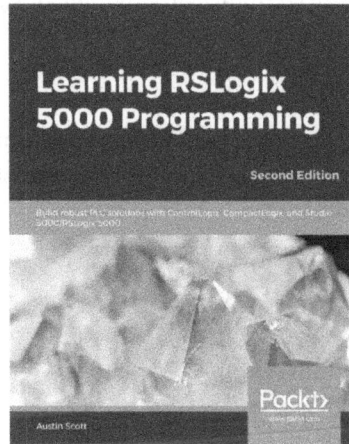

Learning RSLogix 5000 Programming - Second Edition

Austin Scott

ISBN: 9781789532463

- Gain insights into Rockwell Automation and the evolution of the Logix platform
- Find out the key platform changes in Studio 5000 and Logix Designer
- Explore a variety of ControlLogix and CompactLogix controllers
- Understand the Rockwell Automation industrial networking fundamentals
- Implement cybersecurity best practices using Rockwell Automation technologies
- Discover the key considerations for engineering a Rockwell Automation solution

Packt is searching for authors like you

If you're interested in becoming an author for Packt, please visit `authors.packtpub.com` and apply today. We have worked with thousands of developers and tech professionals, just like you, to help them share their insight with the global tech community. You can make a general application, apply for a specific hot topic that we are recruiting an author for, or submit your own idea.

Share your thoughts

Now you've finished *The Art of Manufacturing*, we'd love to hear your thoughts! Scan the QR code below to go straight to the Amazon review page for this book and share your feedback or leave a review on the site that you purchased it from.

`https://packt.link/r/1804619450`

Your review is important to us and the tech community and will help us make sure we're delivering excellent quality content.

Download a free PDF copy of this book

Thanks for purchasing this book!

Do you like to read on the go but are unable to carry your print books everywhere?

Is your eBook purchase not compatible with the device of your choice?

Don't worry, now with every Packt book you get a DRM-free PDF version of that book at no cost.

Read anywhere, any place, on any device. Search, copy, and paste code from your favorite technical books directly into your application.

The perks don't stop there, you can get exclusive access to discounts, newsletters, and great free content in your inbox daily!

Follow these simple steps to get the benefits:

1. Scan the QR code or visit the link below:

https://packt.link/free-ebook/9781804619452

2. Submit your proof of purchase

That's it! We'll send your free PDF and other benefits to your email directly.

www.ingramcontent.com/pod-product-compliance
Lightning Source LLC
Chambersburg PA
CBHW080551220326
41599CB00032B/6435